Communications in Computer and Information Science   784

Commenced Publication in 2007
Founding and Former Series Editors:
Alfredo Cuzzocrea, Xiaoyong Du, Orhun Kara, Ting Liu, Dominik Ślęzak,
and Xiaokang Yang

Editorial Board

More information about this series at http://www.springer.com/series/7899

Juanzi Li · Ming Zhou
Guilin Qi · Ni Lao
Tong Ruan · Jianfeng Du (Eds.)

Knowledge Graph and Semantic Computing

Language, Knowledge, and Intelligence

Second China Conference, CCKS 2017
Chengdu, China, August 26–29, 2017
Revised Selected Papers

 Springer

Editors
Juanzi Li
Tsinghua University
Beijing
China

Ming Zhou
Beijing Xigema Center
Beijing
China

Guilin Qi
School of Computer Science
and Engineering
Southeast University
Nanjing, Jiangsu
China

Ni Lao
Google
Mountain View, CA
USA

Tong Ruan
East China University of Science
and Technology
Shanghai
China

Jianfeng Du
Guangdong University of Foreign Studies
Guangzhou
China

ISSN 1865-0929 ISSN 1865-0937 (electronic)
Communications in Computer and Information Science
ISBN 978-981-10-7358-8 ISBN 978-981-10-7359-5 (eBook)
https://doi.org/10.1007/978-981-10-7359-5

Library of Congress Control Number: 2017964215

Printed on acid-free paper

This Springer imprint is published by Springer Nature
The registered company is Springer Nature Singapore Pte Ltd.
The registered company address is: 152 Beach Road, #21-01/04 Gateway East, Singapore 189721, Singapore

Preface

This volume contains the papers presented at CCKS 2017: the China Conference on Knowledge Graph and Semantic Computing held during August 26–29, 2017, in Chengdu.

CCKS is organized by the Technical Committee on Language and Knowledge Computing of the Chinese Information Processing Society (CIPS). CCKS 2017 was the second edition of the conference series. CCKS 2016 was the merger of two premier relevant forums held previously: the Chinese Knowledge Graph Symposium (KGS) and the Chinese Semantic Web and Web Science Conference (CSWS). KGS was held in Beijing in 2013, in Nanjing in 2014, and in Yichang in 2015. CSWS was first held in Beijing in 2006, and has been the main forum for research on the Semantic (Web) technologies in China for nearly ten years. CCKS covers wide research fields including knowledge graph, the Semantic Web, linked data, NLP, knowledge representation, graph databases etc. It aims to be the top forum on knowledge graph and semantic technologies for Chinese researchers and practitioners from academia, industry, and government.

The theme of this year was "Language, Knowledge, and Intelligence."

There were 85 submissions. Each submission was reviewed by at least three Program Committee members. The committee decided to accept 19 full papers (including 11 papers written in English and eight papers written in Chinese) and six short papers. The program also included four invited keynotes, five tutorials, one panel, and one industrial forum. The CCKS volume contains revised versions of 11 full papers and six short papers. This year's invited talks were given by Prof. Amit Sheth from Wright State University, Prof. Hans Uszkoreit from DFKI, Prof. Jun Zhao from the Chinese Academy of Sciences, and Dr. Chin-Yew Lin from Microsoft.

The hard work and close collaboration of a number of people contributed to the success of this conference. We would like to thank the members of the Organizing Committee and Program Committee for their support as well as the authors and participants, who are the primary reason for the success of this conference.

Finally, we appreciate the sponsorships from Gridsum, ChinaScope, and Classic Law Institute as platinum sponsors, Summba, dfx, and Unisound as gold sponsors, Credit Harmony Research as the silver sponsor, and NewaSoft and Lewei Tech as bronze sponsors.

August 2017

Juanzi Li
Ming Zhou
Guilin Qi
Ni Lao
Tong Ruan
Jianfeng Du

Organization

CCKS 2017 was organized by the Language and Knowledge of Computing Committee of the Chinese Information Processing Society.

General Chairs

Juanzi Li   Tsinghua University, China
Ming Zhou   Microsoft Research Asia

Program Chairs

Guilin Qi   Southeast University, China
Ni Lao   Google

Advanced Lectures Chairs

Bing Qin   Harbin Institute of Technology, China
Zhiyuan Liu   Tsinghua University, China

Industrial Forum Chairs

Jun Yan   Microsoft Research Asia, China
Dianxia Xie   Shanghai Haizhi Technology Co. Ltd., China

Evaluation Chairs

Yanghua Xiao   Fudan University, China
Zhichun Wang   Beijing Normal University, China

Poster/Demo Chairs

Huajun Chen   Zhejiang University, China
Kang Liu   Institute of Automation, Chinese Academy of Sciences, China

Local Chairs

Yajun Du   Xihua University, China
Yongquan Fan   Xihua University, China

Sponsorship Chairs

Tieyun Qian   Wuhan University, China
Wei Hu   Nanjing University, China

Publication Chairs

Tong Ruan   East China University of Science and Technology, China
Jianfeng Du   Guangdong University of Foreign Studies, China

Publicity Chairs

Haofen Wang   Shenzhen Gowild Intelligent Technology Company, China
Xiaowang Zhang   Tianjin University, China

Top Conference Review Chairs

Quan Wang   Institute of Information Engineering, Chinese Academy
                     of Sciences, China
Gong Cheng   Nanjing University, China

Area Chairs

Knowledge Representation and Reasoning

Qili Zhu   Shanghai Jiaotong University, China
Jianfeng Du   Guangdong University of Foreign Studies, China

Knowledge Graph Construction and Information Extraction

Xipeng Qiu   Fudan University, China
Shizhu He   Institute of Automation, Chinese Academy of Sciences, China

Knowledge Storage and Indexing

Lei Zhou   Beijing University, China
Xin Wang   Tianjin University, China

Language Understanding and Machine Reading

Wenliang Chen   Shuzhou University, China
Jianyi Guo   Kunming University of Science and Technology, China

Question Answering and Semantic Search

Bing Wang   Institute of Information Engineering, Chinese Academy
                     of Sciences, China
Weinan Zhang   Harbin Institute of Technology, China

Linked Data and Semantic Integration

Wei Hu	Nanjing University, China
Peng Wang	Dongnan University, China

Program Committee

Xipeng Qiu	Fudan University, China
Kun Xu	Beijing University, China
Decao Song	Thomson Reuters
Xiaowang Zhang	Tianjin University, China
Ran Yu	L3S
Jianfeng Du	Guangdong University of Foreign Studies, China
Yansong Feng	Beijing University, China
Xin Wang	Tianjin University, China
Zhigang Wang	Tsinghua University, China
Jie Lu	IBM
Yankai Lin	Tsinghua University, China
Quang Wang	Institute of Automation, Chinese Academy of Sciences, China
Gang Wu	Northeastern University, China
Songfeng Huang	IBMfen
Saisai Gong	Nanjing University, China
Chengjie Sun	Harbin Institute of Technology, China
Yu Hong	ShuZhou University, China
Tao Ge	Beijing University, China
Shizhu He	Institute of Automation, Chinese Academy of Sciences, China
Xiaogang Ma	RPI
Lidong Bing	CMU
Yafang Wang	Shangdong University, China
Gerard De Melo	Tsinghua University, China
Gong Cheng	Nanjing University, China
Ruobing Xie	Tsinghua University, China
Zhiyuan Liu	Tsinghua University, China
Shu Guo	Institute of Information Engineering, Chinese Academy of Sciences, China
Bowei Zhou	Rensselaer Polytechnic Institute, USA
Fu Zhang	Northeastern University, China
Bing Qin	Harbin Institute of Technology, China
Yao Meng	Fujitsu Research Development Centre Ltd.
Huajun Chen	Zhejiang University, China
Jeff Pan	University of Aberdeen, UK
Yu Hong	Shuzhou University, China
Guilin Qi	Southeast University, China

Junhu Wang	Griffith University, Australia
Yuanfang Li	Monash University, Australia
Pingpeng Yuan	Huazhong University of Science and Technology, China
Jingwei Cheng	Northeastern University, China
Kang Liu	Institute of Automation, Chinese Academy of Sciences, China
Linlin Wang	Tsinghua University, China
Yixin Cao	Tsinghua University, China
Heng Ji	RPI
Ziqi Zhang	The University of Sheffield, UK
Yuping Shen	Sun Yat-sen University, China
Xianpei Han	Institute of Software, Chinese Academy of Science, China
Bingfeng Luo	Beijing University, China
Liwei Chen	Beijing University, China
Hai Wan	Sun Yat-sen University, China
Zhe Wang	Griffith University, Australia
YuMing Shen	Guangdong University of Foreign Studies, China
Heng Zhang	Huazhong University of Science and Technology, China
Juanzi Li	Tsinghua University, China
Jun Zhao	Chinese Academy of Sciences, China
Ganggao Zhu	Madrid Polytechnic, Spain

Sponsors

九 章 研 究 所
CLASSIC LAW INSTITUTE

SUMMBA

Unisound

信和研究院
CREDIT HARMONY RESEARCH

NewaSoft泥娃

乐为科技
LEWEI TECH

Contents

Knowledge Base Completion by Learning to Rank Model

Yong Huang[1,2] and Zhichun Wang[1,2(✉)]

[1] Beijing Advanced Innovation Center for Future Education,
Beijing Normal University, XinJieKouWai St. 19, HaiDian District, Beijing 100875,
People's Republic of China
yhuang@mail.bnu.edu.cn, zcwang@bnu.edu.cn
[2] College of Information Science and Technology, Beijing Normal University,
XinJieKouWai St. 19, HaiDian District, Beijing 100875, People's Republic of China

Abstract. Knowledge base (KB) completion aims to predict new facts from the existing ones in KBs. There are many KB completion approaches, one of the state-of-art approaches is Path Ranking Algorithm (PRA), which predicts new facts based on path types connecting entities. PRA treats the relation prediction as a classification problem, and logistic regression is used as the classification model. In this work, we consider the relation prediction as a ranking problem; learning to rank model is trained on path types to predict new facts. Experiments on YAGO show that our proposed approach outperforms approaches using classification models.

Keywords: Knowledge base completion · Path ranking
Learning to rank

1 Introduction

Recent years have witnessed a rapid growth of Knowledge Bases (KBs), such as YAGO [13], DBpedia [1], Google Knowledge Vault [4] etc. Usually, large-scale KBs are built automatically by extracting from the web text or other resources. KBs contain large amount of facts about various entities, and they are very useful to many applications such as Question Answering, biomedical information. Despite their large number of facts, they are still incomplete and missing a huge number of facts. To deal with this problem, many works have been done on KB completion, which aim to fill in missing facts by using existed facts to predict unknown facts.

Symbolic approaches use rules or relation paths to infer new facts in a KB. Lao [7] introduced the Path Ranking Algorithm (PRA), which uses random walks to search through bounded length of paths connecting entity pairs of multi-relational instances. These paths are used as features in a classifier that predicts new instances of the given relation. In PRA, each relation path can be viewed as a logic rule, therefore PRA actually is a kind of discriminatively trained logical inference [6].

© Springer Nature Singapore Pte Ltd. 2017
J. Li et al. (Eds.): CCKS 2017, CCIS 784, pp. 1–6, 2017.
https://doi.org/10.1007/978-981-10-7359-5_1

Other KB completion methods such as embedding approaches [2,15] are also used recently. Embedding views KB completion problem as matrix completion, and learn low-dimensional representations of both entities and relations in KB, this can be used to infer new facts. There are also some work trying to combine symbolic technique and embedding technique most recently, including path-based TransE [9] etc.

This paper focuses on PRA and its extensions. PRA extracts relational path features to build classification models using logistic regression. It is based on local close world assumption (LCWA) [4] to generate negative entity pairs. However, in real KB there are too many negative entity pairs that positive and negative triples are extremely imbalance. What's more, KB completion use the candidate entity pairs to fill in the missing facts, ranking these candidates are much more reasonable instead of classifying or scoring these entity pairs. Here we consider a novel KB completion method by learning to rank model, and experiments show that this model is extremely beneficial for inferring new facts.

The article is structured as follows. We introduce our learning to rank model for KB completion in Sect. 2, and we report our experiments in Sect. 3, then review some symbolic related work in Sect. 4. Our conclusion and future works are in Sect. 5

2   Learning to Rank Model

Learning to rank model for KB Completion is a two step process: (1) generating feature matrix, (2) inferring new facts for each relation. In our approach, we follow the same method in PRA to extract relational features of entity pairs. While predicting new facts, we define pairwise objective function, with the purpose of improving entity rank rather than entity pair scores. And we use lambdaMART as a learning to rank model in comparison with PRA-style models, directly minimize mean average precision loss.

Feature Computing. Given a target relation r_i in a KB, we collect a set of entity pairs that have relation r_1, R_i is a set satisfying:

$$R_i = \{(h_{ij}, t_{ij}) | (h_{ij}, r_i, t_{ij}) \in KB\}$$

and then for R_i, we generate negative entity pairs following the Local Close World Assumption. For each entity pair (h_{ij}, t_{ij}), we perform random walk with restart to collect relational path types that connect (h_{ij}, t_{ij}) with a bounded path length. After random walk, we get a set of path types $P_i = \{p_i | (h_{ij}, p_i, t_{ij}) \in KB\}$. P_i collects the existed relational path types for relation r_i. These path types are used as relational features to make prediction of new entity pairs. We binarize these path type values to compute feature vector for each entity pair.

Entity Pairs Ranking. While predicting new facts in a KB, we are inspired by learning to rank approaches [3]. If real entity pairs exist in a KB, their rank should be higher than these fake entity pairs. Hence we should learn entity rank instead of learning scores of entity pairs. We explore different learning to rank methods to build pairwise objective function in this paper. Providing a group of entity pairs, we consider $(h_{ij}, t_{ij}) = 1$ if $h_{ij}, t_{ij} \in KB$ else $(h_{ij}, t_{ij}) = -1$. If there is a query for $(h_{ij}, r_i, ?)$, we generate a list of ranked tail entity, and the positive pairs are considered to better fit the relation while the negative pairs are not.

We take a state-of-art boosting tree method called gradient boosting tree or lambdaMART as our learning to rank approach. This method takes positive and negative entity pairs as a partial order. Instead of minimizing the log loss function of logistic regression, we directly optimize MAP using gradient boosting to learning entity pairs' rank. The output of LambdaMART can be defined as

$$L(r_i|w, c) = \sum_{i=1}^{K} \alpha_i f_i(x) + \sum_{i=1}^{n} l(f(x), f(\hat{x})) + \frac{C}{2} w^T w$$

where in the first part of the formula each $f_i(x)$ is a function modeled by a single tree, α_i is the learned weight with ith tree and K is the number of trees. The second part of is pairwise loss measuring the distance between predicted values and actual values. And the third part is regularization. L2 penalty on leaf values are added to avoid over-fitting. C is constant. Stochastic gradient boosting is used to improve accuracy on each iteration of a base tree.

3 Experiments

We compare our evaluation results of learning to rank methods with that of PRA and SFE [5] in YAGO KB completion task. SFE is another PRA-style KB completion method, it uses subgraph feature extraction to compute much more comprehensive path type features. The baseline model can be found here[1]. Both PRA and learning to rank method use this baseline model to extract path features.

Data. To evaluate our approach, we used the data from YAGO knowledge base. YAGO is built automatically from Wikipedia, GeoNames, and WordNet. Currently, YAGO2 contains 37 kind of relations, more than 10 million entities and more than 120 million facts about these entities. The YAGO2 data can be found here[2]. For each entity pair in KB, we randomly generate eight negative entity pairs using LCWA. For example, an entity pair$(Beijing, isCapitalOf, China)$ can generate a fake entity pair$(Shanghai, isCapitalOf, China)$ and many others. 37 relations are tested in our experiments. There are 124597 entity pairs on average for training, and 21810 for testing.

As we state above, if we use Logistic Regression in PRA and SFE, it's approximated by a learning to rank problem using binary classification. Hence we use LR as PRA and SFE's baseline with python framework[3] to train LR and learning

---

[1] https://github.com/matt-gardner/pra.

[2] http://www.mpi-inf.mpg.de/.

[3] http://scikit-learn.org/stable.

to rank model. We set L2 parameter in LR is a range from [0.01,1]. For learning to rank method we set lambdaMART tree size 1000. We set the same L2 parameter as LR in lambdaMART tree. We set C a default value 1. We present the optimal result as the experimental results in this paper.

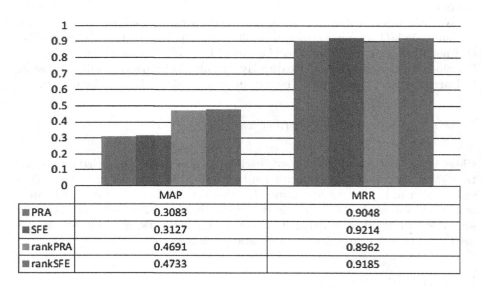

	MAP	MRR
■ PRA	0.3083	0.9048
■ SFE	0.3127	0.9214
■ rankPRA	0.4691	0.8962
■ rankSFE	0.4733	0.9185

Fig. 1. Evaluation results: MAP and MRR

Result. We use mean average precision (MAP) and mean reciprocal rank (MRR) as metrics, which are commonly used in KB completion tasks [5,14]. Both metrics effect evaluate ranking methods, if the true entity pairs rank before the fake entity pairs, the MAP and MRR result will get higher. We compare PRA, SFE with LR model and learning to rank model called rankPRA and rankSFE. The results are presented in Fig. 1. We can conclude that learning entity rank outperforms PRA and SFE significantly on MAP, and get comparable MRR score. In fact, tree methods get 35 kinds of relations' AP score that are higher than PRA and SFE. Only four relations scores are lower than PRA and SFE in RR score. Since tree method is nonlinear, it's difficult to output the learned weight of relational path.

We further explore all relations' AP scores, and results show most relations get significantly promoted than linear model. More than 20 relations' scores are higher than 0.5. Our model shows all relations scores in YAGO2. We can see that most relations get satisfying scores, while some relations get poor score like *imports* and *exports*. We analyze these relations and find triples like (Bangladesh, exports, wordnet_fertilizer_14859344) are difficult for PRA to lookup discriminative path, so we should find better path types in the future work. But for most relations, PRA feature computing can find expressive path feature, and learning to rank methods improve the MAP and MRR score greatly.

4 Related Work

We review symbolic KB completion methods in this section. Symbolic KB completion methods assumes that the existence of an edge can be predicted by extracting features from observed edges in the graph. Predict the triple (Beijing, isLocatedIn, China)from the existence of path (Beijing, hasUniversity, Tsinghua_University, isLocatedIn, China). Other methods like TransE etc. low-dimensional representations approaches are excluded in this paper.

Inductive Logic Programming and Rule Mining. Inductive logic programming (ILP) uses logic programming as a uniform representation to infer facts. Rule mining methods such as Wang [16] learn rules by mining frequent predicate cycles in KB. ILP [10] was Firstly introduced in 1991, another ILP system named First Order Inductive Logic (FOIL) was proposed by Quinlan [12]. It constructs horn clause programs from training examples. For example, isMarriedTo(a,b) $\wedge$ hasChild(a,c) $\Rightarrow$ hasChild(b,c) can be used to infer the fact that $hasChild(DonaldTrump, IvankaTrump)$, if we have learned the rule with variables a, b, c bound to IvanaMarieTrump, DonaldTrump, IvankaTrump. These systems use open world assumption and are easily interpretable, owing to the learned rules only cover a subset of patterns.

Path Ranking Algorithm. PRA [8] uses random walks through bounded length of paths to predict links in multi-relational graphs, and discovers paths by enumerating each relation's entity pairs. The key idea of PRA is using path probabilities as features to build classification methods and predicting hidden relations. PRA can get much more precisely path types compared to ILP, and get comparable performance to embedding methods [11].

PRA Extensions. Many other symbolic methods have been proposed. Subgraph Feature Extraction (SFE) proposed by Gardner [5] is another PRA-style approach. Instead of computing path type probabilities, SFE binarize it as features. This approach can not only be more efficient, but also gain more expressive path features [6] leading better performance. Wang [14] proposed a multi-task learning strategy to cluster some highly correlated relation before using PRA, referred as CPRA. CPRA can effectively identify coherent clusters whose relations are highly correlated.

5 Conclusion

This paper proposes a novel approach to solve KBs completion problem. We follow PRA and apply learning to rank method to rank entity pairs in YAGO. Experiments show that our approach performs better than PRA and SFE. More expressive features can be added to PRA feature computing in the future work, and we will explore new ranking methods to improve symbolic KB completion performance.

Acknowledgement. The work is supported by project of Beijing Advanced Innovation Center for Future Education (BJAICFE2016IR-002).

References

1. Bizer, C., Lehmann, J., Kobilarov, G., Auer, S., Becker, C., Cyganiak, R., Hellmann, S.: Dbpedia-a crystallization point for the web of data. Web Semant. Sci. Serv. Agents World Wide Web **7**(3), 154–165 (2009)
2. Bordes, A., Usunier, N., Garcia-Duran, A., Weston, J., Yakhnenko, O.: Translating embeddings for modeling multi-relational data. In: Advances in Neural Information Processing Systems, pp. 2787–2795 (2013)
3. Cao, Z., Qin, T., Liu, T.-Y., Tsai, M.-F., Li, H.: Learning to rank: from pairwise approach to listwise approach. In: Proceedings of the 24th International Conference on Machine Learning, pp. 129–136. ACM (2007)
4. Dong, X., Gabrilovich, E., Heitz, G., Horn, W., Lao, N., Murphy, K., Strohmann, T., Sun, S., Zhang, W.: Knowledge vault: a web-scale approach to probabilistic knowledge fusion. In: Proceedings of the 20th ACM SIGKDD International Conference on Knowledge Discovery and Data Mining, pp. 601–610. ACM (2014)
5. Gardner, M., Mitchell, T.: Efficient and expressive knowledge base completion using subgraph feature extraction. In: Proceedings of the 2015 Conference on Empirical Methods in Natural Language Processing, pp. 1488–1498 (2015)
6. Gardner, M., Talukdar, P.P., Krishnamurthy, J., Mitchell, T.: Incorporating vector space similarity in random walk inference over knowledge bases. In: Proceedings of the 2014 Conference on Empirical Methods in Natural Language Processing (2014)
7. Lao, N., Cohen, W.W.: Relational retrieval using a combination of path-constrained random walks. Mach. Learn. **81**, 53–67 (2010)
8. Lao, N., Mitchell, T., Cohen, W.W.: Random walk inference and learning in a large scale knowledge base. In: Proceedings of the Conference on Empirical Methods in Natural Language Processing, EMNLP 2011, pp. 529–539. Association for Computational Linguistics, Stroudsburg (2011)
9. Lin, Y., Liu, Z., Luan, H.-B., Sun, M., Rao, S., Liu, S.: Modeling relation paths for representation learning of knowledge bases. In: Proceedings of the Conference on Empirical Methods in Natural Language Processing (EMNLP 2015) (2015)
10. Muggleton, S.: Inductive logic programming. New Gener. Comput. **8**(4), 295–318 (1991)
11. Nickel, M., Murphy, K., Tresp, V., Gabrilovich, E.: A review of relational machine learning for knowledge graphs. Proc. IEEE **104**, 11–33 (2016)
12. Quinlan, J.R., Cameron-Jones, R.M.: FOIL: a midterm report. In: Brazdil, P.B. (ed.) ECML 1993. LNCS, vol. 667, pp. 1–20. Springer, Heidelberg (1993). https://doi.org/10.1007/3-540-56602-3_124
13. Suchanek, F.M., Kasneci, G., Weikum, G.: Yago: a large ontology from wikipedia and wordnet. Web Semant. Sci. Serv. Agents World Wide Web **6**(3), 203–217 (2008)
14. Wang, Q., Liu, J., Luo, Y., Wang, B., Lin, C.-Y.: Knowledge base completion via coupled path ranking. In: ACL (2016)
15. Wang, Z., Zhang, J., Feng, J., Chen, Z.: Knowledge graph embedding by translating on hyperplanes. In: Proceedings of the 28th AAAI Conference on Artificial Intelligence (AAAI 2014), pp. 1112–1119 (2014)
16. Wang, Z., Li, J.-Z.: Rdf2rules: learning rules from RDF knowledge bases by mining frequent predicate cycles. CoRR, abs/1512.07734 (2015)

Path-Based Learning for Plant Domain Knowledge Graph

Cuicui Dong[1], Huifang Du[1], Yaru Du[1], Ying Chen[1], Wenzhe Li[2], and Ming Zhao[1(✉)]

[1] College of Information and Electrical Engineering, China Agricultural University,
Beijing 100083, China
zhaoming@cau.edu.cn
[2] University of Southern California, Los Angeles, USA

Abstract. Learning to embed the knowledge graph has been a hot topic in research communities. As for that, TransE is a promising method that can achieve state-of-art performance for many of the benchmark tasks. However, none of the previous work considers the knowledge graph in plant domain in which case the properties of the graph are significantly different. For the knowledge graph in plant domain, most of its relations belong to one-to-many, many-to-one or many-to-many types (actually majority of them are attribute-type relations), which are not in the scope of consideration for classical TransE model. In order to deal with such unique challenges, we propose a novel model called PTA (path-based TransE for attributes). It constructs the relation path by combining attributes and hyponymy relations, and embeds them to a lower dimensional space as well. We conduct extensive experiments on link prediction task where the performance is measured by mean rank and Hit@10. The results show that our new model significantly outperforms other competing methods on several different tasks.

Keywords: Knowledge graph · PTA · TransE · PtransE

1 Introduction

A knowledge graph is a multi-relational graph [1, 2] composed of nodes and edges where each node represents an entity and each edge represents a relation. Knowledge graph such as Freebase [3] and WordNet [4, 5] are the great tools to support many AI-related tasks such as question-answering [6, 7], chatbot [8, 9], search [10, 11], etc. In addition to the open-domain knowledge graph, there is growing interest in building domain-specific ones such as for biomedical [12, 13], news, film and television applications.

One important but challenging task for knowledge graph is reasoning [14, 15], which plays the key role for link prediction [16, 17] or knowledge completion [18, 19]. The traditional methods of reasoning mainly use the symbolic systems and formal logics [20, 21]. However, the reasoning by formal logic has some major drawbacks: (i) it is not tractable to deal with long-range reasoning over very large knowledge graph. (ii) it performs poorly when the graph is sparse which is common in practice. As with the increasing popularity of deep learning, the community begins to seek another way of representing knowledge graph so called distributed representation [22–24]. In this case, we embed the knowledge graph into a lower dimensional space such that each entity and relation can be represented as a fixed-size continuous vector, while the graph also

© Springer Nature Singapore Pte Ltd. 2017
J. Li et al. (Eds.): CCKS 2017, CCIS 784, pp. 7–17, 2017.
https://doi.org/10.1007/978-981-10-7359-5_2

preserves certain properties of original one. Although this approach provides a reasonable solution for the problems above, learning the embedding vectors for entities and relations is still challenging.

In general, embedding technique represents each entity or relation within a triple (h, r, t) as a K-dimensional vector and defines a score function (h, r, t) to measure the plausibility of this triple in the embedding space. Because of the effectiveness of this scheme, there has been a lot of work follow this line of research. Some of the important work includes structured embedding [25], semantic matching energy [2, 26], latent factor model [11], neural network model [19], matrix decomposition model [27] and translation-based model (TransE) [10], etc.

Among all of them, the TransE model gains particular popularity among the research community, not only because TransE model provides simplicity but also effectiveness and efficiency. In TransE model, we represent a relation by a translation vector r such that the head and tail entities in a triple (h, r, t) can be connected by r with low error. On the other hand, for triples those who do not belong to the knowledge graph, we expect higher error. The whole framework is to minimize the margin loss [28]. Even though TransE model achieves state-of-art performance on many benchmark tasks, it does have some major drawbacks.i.e. it cannot handle beyond one-to-one type relations.

In this paper, we consider the knowledge graph in plant domain whose properties are significantly different from previously studied ones. In plant domain, there are lots of non-taxonomic relations, where only few of them are denoting between entities (i.e. crop rotation, inter-cropping, etc.), while the rest of them are for attribute-type relations such as disease, pests, nutritional value, medicinal value, growth environment, cultivation techniques and etc. More importantly, each entity may have multiple attribute values, and vice versa. Because of that, most relations in plant domain belong to the types of one-to-many, many-to-one and many-to-many, which make the problem even harder.

In order to handle such unique challenges, we propose a new model called PTA (*path-based TransE for attributes*). It constructs the relation path by combining attributes and hyponymy relations, and embeds them to a lower dimensional vector space. The intuition comes from the translation-based operation and can effectively solve for long-range reasoning. In fact, some of the previous models such as transH [29] are also proposed to handle such cases. However, our method is inherently different from others and is more suitable for knowledge graph having lots of attribute-type relations. We conduct extensive experiments on link prediction task where the performance are measured by mean rank and Hit@10. The results show that our new model significantly outperforms competing methods on several different tasks.

Finally, the main contributions of this work can be summarized as:

As far as we know, this is the first work that considers plant domain knowledge graph whose properties are significantly different from previously studied ones. Because of that, our work has great potential to raise the attention from this community to and let them realize the value of the knowledge graph techniques.

We propose a novel model to deal with one-to-many, many-to-one and many-to-many relation types. Although we mainly discuss plant domain knowledge graph, our approach can be easily generalized to other domains as well.

The experiments show that our new model significantly outperforms other competing methods. This again gives the indication that path-based learning scheme may benefit further findings as well.

2 Approach

We will first take a look at the TranE model which acts as the building block for our new model. Then we provide more details about PTA model.

2.1 TransE Model

TransE is a translation-based model for representation learning. The intuition comes from the semantic translation invariance between word embeddings. The model considers the relation in the knowledge graph as a translation vector between head entity and tail entity. Because that each entity maps to a low-dimensional space, it greatly reduces the number of model parameters compared to the previously proposed models, and it also simplies the computation.i.e. given a triple, (tomato, disease_is, strip_rot), when we get the vector representation of head entity - tomato, and relation - disease is, we can simply compute the representation for tail entity - strip_rot.

Given a dataset S consisting of triples, E denotes the entity set, R denotes the relation set, then we have $h, t \in E, r \in R$. For any triples (h, r, t) in set S, TransE model tries to keep the algebraic operation such that $(h + r \approx t)$, which is shown in Fig. 1.

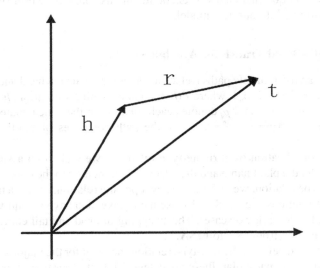

Fig. 1. Illustration of TransE Model

2.2 PTA Model

As a variation of neural network, TransE model [2] maps the entities and relations in the knowledge graph into an embedding space where we assume the basic algebraic operations hold (*i.e.* $h + r \approx t$). Although it is shown that the model performs reasonably well in many benchmark tasks [1], it is unable to deal with one-to-many, many-to-one and many-to-many relations. Suppose we have an one-to-many relation called disease is, then given two triples with the same head entities and relations, (tomatoes, disease_is, strip rot) and (tomatoes, disease_is, bacterial leaf spot), the tail entities tend to map to the same point in the embedding space which does not agree with the fact that these two are actually semantically different. By considering the unique properties of the plant domain, we propose a novel method called PTA to solve this problem.

In the plant domain, there are a lot of non-taxonomic relations. However, only few of them are denoting between entities (i.e. crop rotation, inter-cropping etc.), while the rest of them are for attribute-type relations such as disease, pests, nutritional value, medicinal value, growth environment, cultivation techniques, etc. More importantly, each entity may have multiple attribute values, and vice versa.

Let's first look at how multiple paths are translated on the knowledge graph as it builds the basis for our new model. Suppose we have these three triples: (Johnson, born in city, San Francisco), (San Francisco, belongs to, California) and (California, is state of, *United* States). By simple reasoning over multiple paths, we can generate a new triple: (Johnson, born in country, United States). In the plant domain, we don't have too many types of entities but instead there are lots of attribute-type relations. In order to efficiently handle unique challenges we encounter in this domain, we bring the idea from path-based reasoning into our new model.

2.3 PTA (Path-Based TransE for Attributes)

In general, we can use multiple paths (relations) to concatenate the head and tail entities such that $P(h, t) = \{p_1, \ldots, p_n\}$, where $P(h, t)$ contains path information, h for the head entity, t for the tail entity, and p_i denotes each relation on the concatenated paths. We can also write it as $h \xrightarrow{r_1}, \ldots, \xrightarrow{r_l} t$, where the path composes of multiple relations, $r_1, \ldots, r_l$.

Given the complication of attribute-type relations, we make two assumptions: (i) within each triple, the plant name and the attribute both belong to the entity type. (ii) for each attribute-type relation, we introduce a new opposite relation. i.e. for a triple (barley diseacludes the relations sucse, ISA, disease name), we also have an equivalent triple, (disease name, FISA, barley disease). The swapping of head and tail entities results in the change of relation from **ISA** to **FISA**.

As we mentioned before, attribute-type relations account for the majority of relations in the plant domain. In particular, these are categorized into four main types.

One-to-One. It mainly includes the relations such as Latin name, English name, etc., and it is usually expressed as a first-order path, $h \xrightarrow{r_1} t$, where r_1 denotes the corresponding attribute name. For instance, potato has a Latin name of Solanum tuberosum, both of which are unique to each other.

One-to-Many. The relations such as alias, category are the typical examples for one-to-many type. It usually contains more than two attribute values, but only relate to one particular type of plant entity. The relation can be expressed as a second-order path $h \xrightarrow{r_2} e \xrightarrow{fisa_i} t$, where r_2 denotes the attribute-type relation. i.e. the alias of tomato include tomatoes, persimmon, etc.

Many-to-One. As opposed to one-to-many relations, there are multiple plant entities relate to one single plant entity. Some of them are boundary, branch, category, medicinal property, medicinal smell, flowering date and etc. Similarly, the relation can be expressed as a second-order path: $h \xrightarrow{isa_i} e \xrightarrow{r_3} t$, where r_3 is the attribute-type relation. i.e. radish and carrots both belong to root plants.

Many-to-Many. It mainly includes the relations such as sowing methods, cultivation techniques, disease relationships, pest relationships, nutritional value, medicinal value, edible value and etc. In this case, the relation can be expressed as a third-order path: $h \xrightarrow{isa_i} e_1 \xrightarrow{r_i} e_2 \xrightarrow{fisa_i} t$, where r_4 denotes the attribute-type relation. i.e. medicinal value of burdock and wasabi include anti-cancer and anti-aging functions.

Figure 2 shows these four cases. We use solid circle to represent the existing entities in the knowledge graph, and use the hollow one to represent newly added fuzzy entities. As shown in the figure, the triples (tomato, alias, tomato) and (tomato, alias, Fan persimmon) can be converted into (Alias attribute, FISA, Tomato) and (alias attribute, FISA, Fan persimmon), where alias attribute is treated as a fuzzy entity.

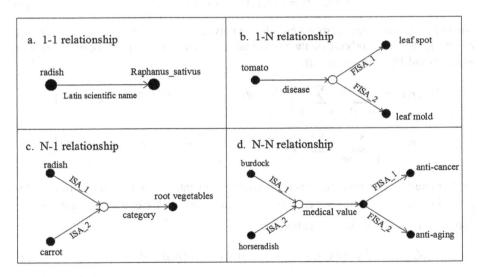

Fig. 2. Four different relation types and its corresponding examples.

2.4 Model Formulation

Similar to TransE model, we use $E(h, p, t)$ to denote the energy for triple (h, p, t). The PTA model also needs to vectorize the relation path by mapping the entities and relations into the same embedding space. Please note that in the relation path $h \xrightarrow{isa_i} e \xrightarrow{r_i} t$, e is a vector for added fuzzy entity, and can be obtained by averaging over several related entities in the triples. In particular, for triples $(h, \mathrm{r}, t_1), \dots, (h, \mathrm{r}, t_n)$, the vector for entity e is defined as: $e = \frac{1}{n} \sum_{i=1}^{n} t_i$, where n is the number of instances. In addition, the vector for relation path is computed by summing over all the relational vectors. i.e. for $p = (r_1, r_2)$, the vector for p is the addition of r_1 and r_2 such that $p = r_1 + r_2$.

For TransE model, when there exists (h, r, t), then we assume $h + r$ and t have semantically similar meanings. The semantic distance is computed as:

$$D(h + r, t) = |h + r - t|_{L1/L2} \tag{1}$$

Here, we can either use L1 or L2 distance.

Similarly, for PTA model, we can define the distance for each triple as:

$$D(h + p, t) = |h + p - t|_{L1/L2} \tag{2}$$

where p is the vector representation for relation path. For triples having many-to-one relation, each path p_i is defined as: $p_i = r_{ISA_i} + r$. On the other hand, for one-to-many relation, we have $p_i = r + r_{FISA_i}$.

For the PTA model, the energy function for each triple is defined as:

$$G(h, r, t) = E(h, r, t) + E(h, p, t) \tag{3}$$

where $E(h, r, t)$ denotes the direct correlation between entities, and $E(h, p, t)$ denotes the energy for the triple relying on the vectorization of relation path p. The computation of $E(h, r, t)$ and $E(h, r, t)$ are as follows:

$$E(h, r, t) = \sum_{(h, r_{isa}, t) \in S} \sum_{(h', r_{isa}, t')} \sum_{i \in N} \left[\delta + d(h_i + r_{isa_i} + r, t) - d(h'_i, r_{isa_i} + r, t') \right]_+$$

$$E(h, p, t) = \sum_{(h, r_{isa}, t) \in S} \sum_{(h', r_{isa}, t')} \sum_{i \in N} \left[\delta + d(h + r_{isa_i} + r_i, t) - d(h', r_{isa_i} + r, t'_i) \right]_+$$

For one-to-many relation, N is the number of tail entities, and for many-to-one relation, N is the number of head entities, $[.]_+$ is the operation for extracting only positive part, and 0 otherwise. $S'_{(h,p,t)}$ is defined as:

$$S'_{(h,p,t)} = (h', p, t)|h' \in E \cup (h, p', t)|p' \in P \cup (h, p, t')|t' \in E \tag{4}$$

Gradient-based method is used to optimize the model parameters, and the process is quite similar as in TransE model.

3 Experiments

In this section, we provide detailed experimental setup and results followed by some discussions.

3.1 Setup

Learning for TransE. We use TransE model to map entities and relations into an embedding space. In plant domain, we select 12 different vegetables types as training data, edible fungi (37 types in total) for testing data, and sprouts (25 types in total) as validation data. The detailed dataset description is shown in Table 1.

Table 1. TransE model dataset

Data set	Size of entities	Size of relations	Size of training set	Size of validation set	Size of test set	Size of total entries
Plants	8780	187	7550	1850	1250	10650

The hyperparameter setup is as follows: the vector size is chosen from $\{50:100\}$; we set the learning rate λ to one of these values – $\{0.001, 0.01, 0.1\}$, and the margin γ to $\{1, 2, 10\}$. Finally, the optimal parameter combination is given by size = 100, $\lambda = 0.001$, $\gamma = 1$.

Learning for PTA. In addition to the mappings of entities to an embedding space, we also need to map the relation path to the same space. Second-order translation model can be used for characterize one-to-many and many-to-one relation types, and third-order translation model for many-to-many relation types. i.e. alias information, the relationship between pests and diseases, nutritional value and medicinal value etc. The data we used to evaluate our new model is shown in Table 2. We randomize the data into training, validation, testing as the ratio of 20: 1: 1.

Table 2. PTA model dataset

Relation types	Size of entities	Size of attribute relation	Size of training set	Size of validation set	Size of test set	Total entries
1 – 1	1764	20	1708	50	50	1808
1 – N	2921	85	3608	170	170	3948
N – 1	2200	39	2984	140	140	3264
N – N	2105	43	2062	136	136	2334
Total	8990	187	10362	496	496	11354

By comparing Tables 1 and 2, we can notice that there are extra 210 entities and 704 triples introduced. We use the same parameter setting for both PTA and TransE models.

3.2 Evaluation

As most of the triples are extracted manually, the quality of the dataset can be guaranteed. However, there are some factors that might affect the coverage of knowledge graph which is our on-going work to dynamically update and fullfill the knowledge base.

We evaluate the model performance via knowledge completion task also called link prediction. Given two elements within a triple, link prediction tries to recover the third element which is missing. i.e. given a head entity tomato and relation alias, the well-trained model should predict the tail entity as either tomatoes or persimmon.

We propose two evaluation metrics, mean rank and *Hits@10*. The mean rank is to measure the average rank of the entity to be predicted, and the *hit@10* is to measure the probability how often the correct entity is in the top 10 list.

Task 1: link prediction without considering relation categories. Table 3 shows the link prediction performance for PTA and TransE model in small-domain plant graph.

Table 3. Effects without considering relation categories

Link prediction evaluation index	Mean rank		Hits@10(%)	
	Raw	Filter	Raw	Filter
TransE	79.1	78.7	28.4	28.8
PTA(ADD)	74.8	72.6	32.8	34.3

As we can see from the result, the PTA model outperforms TransE model in terms of both mean rank and hit@10 criteria. The value of mean rank decreases to 72.6 and the value of Hits@10 is increased to 34.3%, which gives about 5% improvement. The result implies that the path-based learning scheme provides good complement to the representation learning of knowledge graph.

Task 2: link prediction by considering relation categories. We construct the relation path using attribute-type relation and ISA relation, to solve one-to-many, many-to-one and many-to-many problems. Table 4 shows the comparison results measured by Hit@10 value on four different relation types, among TransE, PTransE and PTA models.

Table 4. Effects with different relation categories

Evaluation index	Left Head entity prediction				Right Tail entity prediction				
	$1-1$	$1-N$	$N-1$	$N-N$	$1-1$	$1-N$	$N-1$	$N-N$	Total
TransE	30.5	25.2	20.6	25.4	31.9	20.1	25.3	24.0	25.4
PTransE(ADD)	32.0	34.9	24.3	27.7	32.7	25.5	35.4	26.4	29.8
PTA(ADD)	32.3	36.1	26.5	31.8	33.5	27.0	38.3	30.4	32.0

We first divide the dataset by category which results in different number of triples belong to one-to-one, one-to-many, many-to-one, many-to-many case, respectively. Because of that, we are unable to compare the mean rank value. Table 4 shows the prediction performance measured by Hit@10 value on different types of datasets, among TransE, PTransE and PTA models.

First consider the correlation between relation category and direction of prediction. The results from Table 4 can be divided into four categories. Because of the symmetric property of one-to-one, many-to-many relations, there is no big difference in terms of performance for the task of predicting head entity and tail entities. On the other hand, we can see the opposite results for prediction of one-to-many and many-to-one relations. i.e. the PTA model gives the predictive accuracy of 36.1% and 38.3% for the task of predicting head entity for one-to-many relation, and tail entity for many-to-one relation, respectively. The results are similar and give the best performance among others. However, for the task of predicting head entity for many-to-one relation and tail entity for one-to-many relation, we get the accuracy as low as 25.5% and 26.0%, which gives the worst performance. We can see why by showing the example of predicting for head entity. In this case, the expansion of relation path opposes to the direction of prediction such that each entity maps to multiple attributes which further increases the prediction accuracy of that entity. On the other hand, for the case of many-to-one, multiple entities map to a single attribute value, which results in performance decrease.

Lastly, we evaluate the predictive performance of different models on the same type of relation.

PTA vs TransE: as we can see from Table 4, PTA model improves the Hit@10 value by 6.6%. In addition, it shows the 13% improvements on predicting tail entity for many-to-one relation.

PTA vs PTransE: they both give improved performance on some complicated tasks. But because of the fact that PTransE mainly uses the second-order reasoning, it is more suitable for large knowledge graph with dense connections and complicated relations. The results also show that PTA model improves the Hit@10 value over PTransE by 2.2%, and 4% for many-to-many case.

4 Conclusion

In this paper, we consider the construction and reasoning over the plant domain knowledge graph. Because that most of the relations are attribute-type, there are significant number of relations belong to one-to-many, many-to-one or many-to-many types, which is inherently different from graph from other domains. In order to handle such special challenges, we propose a novel method called PTA which constructs the relation path by combining attributes and hyponymy relations, and embeds them to a lower dimensional space as well. The experimental results show its superior performance over other competing methods.

References

1. Nickel, M., Murphy, K., Tresp, V., Gabrilovich, E.: A review of relational machine learning for knowledge graphs. arXiv preprint arXiv:1503.00759 (2015)
2. Bordes, A., Usunier, N., Garcia-Duran, A., Weston, J., Yakhnenko, O.: Translating embeddings for modeling multi-relational data. In: Advances in Neural Information Processing Systems, pp. 2787–2795 (2013)

3. Bollacker, K., Evans, C., Paritosh, P., Sturge, T., Taylor, J.: Freebase: a collaboratively created graph database for structuring human knowledge. In: Proceedings of the 2008 ACM SIGMOD International Conference on Management of Data, pp. 1247–1250. ACM, June 2008

4. Miller, G.A.: WordNet: a lexical database for English. Commun. ACM **38**(11), 39–41 (1995)

5. Miller, G.A., Beckwith, R., Fellbaum, C., Gross, D., Miller, K.J.: Introduction to WordNet: an on-line lexical database. Int. J. Lexicogr. **3**(4), 235–244 (1990)

6. Yao, X., Van Durme, B.: Information extraction over structured data: question answering with freebase. In: ACL (1), pp. 956–966 (2014)

7. Frank, A., Krieger, H.U., Xu, F., Uszkoreit, H., Crysmann, B., Jrg, B., Schfer, U.: Question answering from structured knowledge sources. J. Appl. Logic **5**(1), 20–48 (2007)

8. Tarau, P., Figa, E.: Knowledge-based conversational agents and virtual storytelling. In: Proceedings of the 2004 ACM Symposium on Applied Computing, pp. 39–44. ACM, March 2004

9. Hakkani-Tr, D., Celikyilmaz, A., Heck, L.P., Tr, G., Zweig, G.: Probabilistic enrichment of knowledge graph entities for relation detection in conversational understanding. In: INTERSPEECH, pp. 2113–2117, September 2014

10. Milne, D.N., Witten, I.H., Nichols, D.M.: A knowledge-based search engine powered by wikipedia. In: Proceedings of the Sixteenth ACM Conference on Information and Knowledge Management, pp. 445–454. ACM, November 2007

11. Matsuo, Y., Sakaki, T., Uchiyama, K., Ishizuka, M.: Graph-based word clustering using a web search engine. In: Proceedings of the 2006 Conference on Empirical Methods in Natural Language Processing, pp. 542–550. Association for Computational Linguistics, July 2006

12. De Bruijn, B., Martin, J.: Getting to the (c)ore of knowledge: mining biomedical literature. Int. J. Med. Informat. **67**(1), 7–18 (2002)

13. Rubin, D.L., Lewis, S.E., Mungall, C.J., Misra, S., Westerfield, M., Ashburner, M., Day-Richter, J., et al.: National center for biomedical ontology: advancing biomedicine through structured organization of scientific knowledge. OMICS: J. Integr. Biol. **10**(2), 185–198 (2006)

14. Fagin, R., Halpern, J.Y., Moses, Y., Vardi, M.: Reasoning About Knowledge. MIT Press, Cambridge (2004)

15. Chein, M., Mugnier, M.L.: Graph-Based Knowledge Representation: Computational Foundations of Conceptual Graphs. Springer, London (2008)

16. LibenNowell, D., Kleinberg, J.: The linkprediction problem for social networks. J. Am. Soc. Inform. Sci. Technol. **58**(7), 1019–1031 (2007)

17. Kunegis, J., Lommatzsch, A.: Learning spectral graph transformations for link prediction. In: Proceedings of the 26th Annual International Conference on Machine Learning, pp. 561–568. ACM, June 2009

18. Lin, Y., Liu, Z., Sun, M., Liu, Y., Zhu, X.: Learning entity and relation embeddings for knowledge graph completion. In: AAAI, pp. 2181–2187, January 2015

19. Socher, R., Chen, D., Manning, C.D., Ng, A.: Reasoning with neural tensor networks for knowledge base completion. In: Advances in Neural Information Processing Systems, pp. 926–934 (2013)

20. Sowa, J.: Knowledge Representation: Logical, Philosophical, and Computational Foundations. PWS Publishing Company, Boston (2000). Book in preparation

21. Moore, R.C.: The role of logic in knowledge representation and common-sense reasoning. SRI International. Artificial Intelligence Center (1982)

22. Ishibuchi, H., Nozaki, K., Tanaka, H.: Distributed representation of fuzzy rules and its application to pattern classification. Fuzzy Sets Syst. **52**(1), 21–32 (1992)

23. Grenander, U., Miller, M.I.: Representations of knowledge in complex systems. J. Roy. Stat. Soc.: Ser. B (Methodol.) **56**, 549–603 (1994)
24. Elman, J.L.: Distributed representations, simple recurrent networks, and grammatical structure. Mach. Learn. **7**(2–3), 195–225 (1991)
25. Bordes, A., Weston, J., Collobert, R., Bengio, Y.: Learning structured embeddings of knowledge bases. In: Conference on Artificial Intelligence (No. EPFLCONF-192344) (2011)
26. Jenatton, R., Roux, N.L., Bordes, A., Obozinski, G.R.: A latent factor model for highly multi-relational data. In: Advances in Neural Information Processing Systems, pp. 3167–3175 (2012)
27. Nickel, M., Tresp, V., Kriegel, H.P.: Factorizing yago: scalable machine learning for linked data. In: Proceedings of the 21st International Conference on World Wide Web, pp. 271–280. ACM, April 2012
28. Lin, Y.: A note on margin-based loss functions in classification. Stat. Probab. Lett. **68**(1), 73–82 (2004)
29. Wang, Z., Zhang, J., Feng, J., Chen, Z.: Knowledge graph embedding by translating on hyperplanes. In: AAAI, pp. 1112–1119, July 2014

A Graph-Based Approach to Incremental Classification in OWL 2 QL Ontology

Changlong Wang[1], Xiaowang Zhang[2(✉)], Zhiyong Feng[3], and Hongwu Qin[1]

[1] School of Computer Science and Engineer Technology,
Northwest Normal University, Lanzhou 730070, China
[2] School of Computer Science and Technology,
Tianjin University, Tianjin 300350, China
xiaowangzhang@tju.edu.cn
[3] School of Computer Software, Tianjin University, Tianjin 300350, China

Abstract. In this paper, we propose an incremental reasoning approach to OWL 2 QL ontologies by mapping an ontology to an updatable digraph with maintaining dynamic transitive closure to obtain incremental classification. Firstly, we transform an ontology to an updatable digraph and then propose a procedure of updating ontology digraph for incremental classification later. Secondly, we develop an algorithm to identify affected paths for incremental classification. Finally, we implement our proposed approach in a prototype incR and then evaluate it on widely-used ontologies. The experiments show that our approach leads to performance gain and outperforms the techniques of module-based incremental classification in OWL 2 QL ontologies.

Keywords: OWL 2QL · Ontology
Incremental reasoning · Classification

1 Introduction

Ontologies expressed in the Web Ontology Language (OWL) and its revision OWL 2 play a central role in the development of the Semantic Web [1]. They are also widely used in biomedical information systems and other areas [2,3]. Those systems based on ontology require efficient and robust reasoning services, among which classification is the core service that computes all subsumption relationships inferred in an ontology between predicate names in the ontology signature, i.e., named concepts (a.k.a. classes), and role s (a.k.a. object-properties). Classification of an ontology can be exploited for various tasks, at both design-time and run-time, ranging from ontology navigation and visualization to query answering.

Devising efficient ontology classification methods and implementations is a challenging issue, since classification is in general an expensive computation. Most popular reasoners for Description Logic (DL) ontologies, i.e., OWL ontologies, such as Racer [4], Pellet [5], FACT [6], and HermiT [7], offer highly optimized classification services for expressive DLs. Those reasoners are on based

© Springer Nature Singapore Pte Ltd. 2017
J. Li et al. (Eds.): CCKS 2017, CCIS 784, pp. 18–29, 2017.
https://doi.org/10.1007/978-981-10-7359-5_3

on tableau (hyper-tableau) algorithms by model construction, extensive experimental studies show that such reasoners have reached very good performances through the years. However, they are still not able to efficiently classify very large ontologies, such as the full versions of GALEN [8] or of the FMA [9] ontology.

In the recent years, the three tractable profiles of OWL 2, i.e., OWL 2 EL, OWL 2 RL, and OWL 2 QL [10] have obtained increasing attention due to their favourable computational properties. Some special reasoners were developed for those profiles. Both CEL [11] and ELK [12] are specifically tailored to intensional reasoning over description logics of the EL family, the former is based on completion rule and the later on consequence. The CB reasoner is a consequence based reasoner for the OWL 2 RL. CB can obtain an impressive gain on very large ontologies, such as full GALEN. However, the current implementation of the CB reasoner is rather specific for particular fragments of Horn-SHIQ (and incomplete for the general case).

In order to void repeating the whole reasoning process from cratch after continuous but relatively small modification during ontology evolving, incremental reasoning techniques are proposed to optimized current reasoners. The based-module incremental reasoning method [13] is suitable for the expressive $SROIQ$ ontology, this approach needs to maintains a collection of modules for derived conclusions. The modules consist of axioms in the ontology that entail the respective conclusion, but they are not necessarily minimal. If no axiom in the module was deleted then the entailment is still valid. The method does not require changes to the reasoning algorithm, but still incurs the cost of computing and storing the modules. For the OWL 2 EL profile, the first incremental classification method [14] is proposed and implemented in the CEL system [11], but this method dose not support Deletion of axioms. The fully incremental classification that supports both addition and deletion of axioms in proposed in [15], this approach adapts DRed algorithm [16]. As OWL 2 RL ontologies can be captured by Datalog programs, some existing incremental techniques in the deductive database context have been adapted to deal with RL ontologies [17].

In recent years, OWL 2 QL ontologies have become increasingly important in Ontology-Based Data Access (OBDA) [18–20,25,26], however, the OWL 2 QL has received so far little attention in classification. Recently, classification method specifically tailored to OWL 2 QL has been investigated only in [21], this approach is based on ontology digraph representation and shows excellent performances in classification of ontologies specified in language of DL-Lite family, which are the logical underpinning of OWL 2 QL. However, the approach in [21] is only suitable for static ontology. To the best of our knowledge, there exists no incremental classification approach and implementation tailored to OWL 2 QL ontology. In this paper, we close this gap by mapping an evolving QL ontology to a dynamic direct graph and maintaining its transitive closure to obtain incremental classification.

The remainder of this paper is structured as follows. After a brief introduction to the preliminaries, in Sect. 3 we describes the procedure of mapping an evolving QL 2 ontology to a dynamic direct graph. In Sect. 4, we present a an algorithm for maintaining identifying the affected paths and maintaining the transitive closure of ontology digraph. In Sect. 5 we propose some strategies for optimisation. We perform extensive experiments to evaluate our methods in Sect. 6 and give conclusion of this work in Sect. 7.

2   Preliminaries

In this section, we briefly introduce OWL 2 QL and bigraph.

2.1   OWL 2 QL

We assume the reader to be reasonably familiar with OWL 2 and its profile OWL 2 QL [1,10], which have become the standard ontology language of W3C and are base description logic (DL) [22].

For convenience reasons, we adopt description logic notion rather than OWL syntax in the formal descriptions. Formally, every DL knowledge base is based on three finite sets of signature symbols: a set N_I of individual names, a set N_C of concepts names and a set N_R of role names. Usually these sets are assumed to be fixed for some applications and therefore not mentioned explicitly. In general, A DL knowledge base includes three parts: TBox, RBox, and RBox, where TBox and RBox describe intensional knowledge, in which axioms represent relationship between concept or role descriptions, ABox describes extensional knowledge about individuals. Usually, a DL ontology is constituted by TBox and RBox, i.e., a DL ontology contains the background knowledge related to a give area.

The DL-Lite family [23] which constitutes the logical underpinning of OWL 2 QL is characterized by allowing unlimited use of existential quantifiers. In details, expressions in OWL 2 QL are formed according to the following syntax:

$$B \rightarrow A | \exists Q$$
$$Q \rightarrow P | P^-$$
$$C \rightarrow B | \neg B | \exists Q.A$$
$$R \rightarrow Q | \neg Q$$

Here A and P are symbols in Σ_P denoting respectively an atomic concept and an atomic role; P^- denote the inverse of P; $\exists Q$, also called *unqualified existential role*, denotes the set of objects related to some object by the role Q; the concept $\exists Q.A$, or qualified existential role, denotes the qualified domain of Q with respect to A, i.e., the set of objects that Q relates to some instance of A. In the following, we call B a basic concept and Q a basic role.

An OWL 2 QL ontology O is a finite set of axioms of the form $B \sqsubseteq C$ and $Q \sqsubseteq R$, where the former denote subsumption between concepts, and the latter subsumption between roles. We call positive inclusions axioms of the form $B_1 \sqsubseteq B_2$, $B_1 \sqsubseteq \exists Q.A$, and negative inclusions axioms of the form $B_1 \sqsubseteq \neg B_2$ and $Q_1 \sqsubseteq \neg Q_2$ and $U_1 \sqsubseteq \neg U_2$. Given axiom $\alpha \in O$, the symbols in the α is denoted with $sig(\alpha)$, and the symbols on the left and right hands of α are denoted with sig(LHS(α)) and $sig(RHS(\alpha))$, respectively.

Traditional intensional reasoning tasks with respect to a given ontology are verification of subsumption and satisfiability of concepts, roles. Strictly related to the intensional reasoning tasks is the classification inference service, which we focus on in this paper. Classification allows to structure the terminology of ontology o in the form of a subsumption hierarchy that provides useful information on the connection between different terms, and can be used to speed up other inference services. It can be defined as follow:

Definition 1. *Let O be a satisfiable ontology over $sig(O)$. The subsumption hierarchy of O is the H_O set of inclusion assertions defined as follows: Let S_1 and S_2 be either two concepts and roles in $sig(O)$. If $O \models S_1 \sqsubseteq S_2$ then $S_1 \sqsubseteq S_2$ belongs to H_O.*

2.2 Graph Theory Notions

We use traditional notions in graph theory. A digraph D consists of a non-empty finite vertex set $V(D)$ and edge set $E(D)$. The degree of a vertex v is denoted by $d(v)$. A path p is a sequence $p = v_1, v_2, ..., v_k$ of distinct vertices in $V(D)$, such that for every $v_i, v_{i+1} \in p$, $(v_i, v_{i+1}) \in E(D)$. A path from u to v is denoted with $u \rightsquigarrow v$. The size of path p is represented by the number of edges contained in p. The number of distinct paths $u \rightsquigarrow v$ is denoted by $Paths(u, v)$. We use $TC(D)$ to denote the transitive closure of a digraph D.

2.3 Digraph Representation of OWL QL 2 Ontologies

The idea of digraph representation of an OWL 2 QL ontology is that, given an OWL 2 QL ontology O and its digraph D_O, each vertex $v \in V(D_O)$ represents a basic concept (role) [21]:

Definition 2. *Let O be an OWL 2 QL ontology. The digraph representation of O, denoted by $D_O = (V, E)$, is built in an inductive way:*

1. *for each atomic concept $A \in sig(O)$, V contains the vertex A;*
2. *for each atomic role $P \in sig(O)$, V contains the four vertices P, P^-, $\exists P$, $\exists P^-$;*
3. *for each concept inclusion $B_1 \sqsubseteq B_2 \in O$, E contains the edge (B_1, B_2);*
4. *for each role inclusion $Q_1 \sqsubseteq Q_2 \in O$, E contains the edges the four (Q_1, Q_2), (Q_1^-, Q_2^-), $(\exists Q_1, \exists Q_2)$, and $(\exists Q_1^-, \exists Q_2^-)$;*
5. *for each concept inclusion $B_1 \sqsubseteq \exists Q.A \in O$, V contains the vertices $\exists Q.A$, and E contains the two edges $(B_1, \exists Q.A)$ and $(\exists Q.A, \exists Q)$.*

Let S_1 and S_2 be two atomic concepts (roles), $O \models S_1 \sqsubseteq S_2$ iff $\exists (S_1, S_2) \in E(TC(D_O))$. Please refer to [21] for details. We use $L(\alpha)$ (resp. $R(\alpha)$) to denote the basic concept or basic role on the left (resp. right) sides of α. We also use $d(L(\alpha))$ or $d(R(\alpha))$ to represent degree of the corresponding vertex. For an axiom of the form $Q_1 \sqsubseteq Q_2$, we use $d(Q_1)$ and $d(Q_2)$ to represent the degrees of Q_1 and Q_2, respectively.

3   Mapping Ontologies Direct Graphs

In this section, we describe the procedure of mapping an ontology to a dynamic digraph. When an axiom is added to an ontology, it is easy to update the corresponding digraph. Let O be an ontology and α^+ be an added axiom, if the basic concepts and basic roles in α^+ are already contained in O, we only insert edge(s) into D_O; if α^+ contains new basic concepts and basic roles, we insert vertex and edge(s) according to Definition 2. Axiom deletions are complicated and discussed in details as follows:

From Definition 2, there exist some differences between concept axiom additions and role axiom additions. Let α^- be a deleted axiom, we consider two cases:

Case 1: α^- is a concept inclusion.
  (1) If $d(L(\alpha^-)) > 1$ and $d(R(\alpha^-)) > 1$, then, only the edge $(L(\alpha^-), R(\alpha^-))$ is removed.
  (2) If $d(L(\alpha^-)) > 1$ and $d(R(\alpha^-)) = 1$, then, the edge $(L(\alpha^-), R(\alpha^-))$ and its head are removed.
  (3) If $d(L(\alpha^-)) = 1$ and $d(R(\alpha^-)) > 1$, then, the edge $(L(\alpha^-), R(\alpha^-))$ and its tail are removed.
  (4) If $d(L(\alpha^-))) = 1$ and $d(R(\alpha^-)) = 1$, then, the edge $(L(\alpha^-), R(\alpha^-))$ is removed, both its tail and head are removed.
Case 2: α^- is a role inclusion axiom of the form $Q_1 \sqsubseteq Q_2$.
  (1) If $d(Q_1) > 1$ and $\text{d}(Q_2) > 1$, then, the 4 edges (Q_1, Q_2), (Q_1^-, Q_2^-), $(\exists Q_1, \exists Q_2)$, $(\exists Q_1^-, \exists Q_2^-)$ need to be removed.
  (2) If $d(Q_1) > 1$ and $d(Q_2) = 1$, then, the 4 edges (Q_1, Q_2), (Q_1^-, Q_2^-), $(\exists Q_1, \exists Q_2)$, $(\exists Q_1^-, \exists Q_2^-)$, and the 4 vertices Q_2, Q_2^-, $\exists Q_2$, $\exists Q_2^-$ need to be removed.
  (3) If $d(Q_1) = 1$ and $d(Q_2) > 1$, then, the 4 edges (Q_1, Q_2), (Q_1^-, Q_2^-), $(\exists Q_1, \exists Q_2)$, $(\exists Q_1^-, \exists Q_2^-)$, and 4 vertices Q_1, Q_1^-, $\exists Q_1$, $\exists Q_1^-$ need to be removed.
  (4) If $d(Q_1) = 1$ and $d(Q_2) = 1$, then, the 4 edges (Q_1, Q_2), (Q_1^-, Q_2^-), $(\exists Q_1, \exists Q_2)$, $(\exists Q_1^-, \exists Q_2^-)$, and the 8 vertices Q_1, Q_1^-, $\exists Q_1$, $\exists Q_1^-$, Q_2, Q_2^-, $\exists Q_2$, $\exists Q_2^-$ are removed.

4   Identifying Affected Paths and Updating Transitive Closure

An edge addition (deletion) into (from) digraph may change the number of path $i \rightsquigarrow j$. It is necessary to identify those affected paths and recompute the transitive closure of such affected subdigraph.

Definition 3. *Given a digraph D_O containing vertices i, j, u, v, and path $i \rightsquigarrow j$, if the number of $i \rightsquigarrow j$ changes after an edge (u, v) is inserted (removed) into (from) D_O, the path $i \rightsquigarrow j$ is called affected path.*

When we add (delete) axioms, no cycle is created in the corresponding digraph D_O. From Lemma 3.1 in [24], the following proposition holds:

Proposition 1. *Let $i \rightsquigarrow j$ be an affected path after the edge (u, v) is inserted (or deleted), the number of the paths $i \rightsquigarrow j$ increases (or decreases) by $Paths(i, u) \times Paths(v, j)$.*

In ontology development, a domain engineer often adds (or deletes) a set of axioms related to a certain concept (role). Such modification will lead to a changed set of edges incident to a common vertex and more affected paths in the corresponding digraph.

Proposition 2 *[24]. Let D_O be a digraph and E_v be a updating set of edges incident to a common vertex v, let $\Delta Paths(i, j)$ be the change to $Paths(i, j)$, $\Delta Paths(i, j) \leftarrow Paths(i, v) \times \Delta Paths(v, j) + \Delta Paths(i, v) \times Paths(v, j) + \Delta Paths(i, v) \times \Delta Paths(v, j)$.*

From above discussion, it is feasible to design an algorithm for incremental classification in QL ontologies. Give a digraph D and any two vertices $i, j \in D$, the number of the path $i \rightsquigarrow j$ can be represented by an $n \times n$ adjacency matrix Q such that $Num(i, j) = 0$ if there exists no path from i to j. The transitive closure of a digraph D is presented by an $n \times n$ adjacency matrix M_D^* such that $M_D^*(i, j) = 1$ if $Q(i, j) \neq 0$, else $M_D^*(i, j) = 0$. While updating the digraph, if $Q(i, j)$ increases from 0, $M_D^*(i, j) = 1$, if $Q(i, j)$ decreases to 0, $M_D^*(i, j) = 0$. Algorithm 1 and 2 describe the procedures of updating digraph while inserting and deleting edges, respectively. Both the algorithms consists of three phases: (i) identifying the affected paths (line 3–8); (ii) updating the path matrix (line 9–19); (iii) updating the transitive closure (line 20–22). The arrays *From* and *To* are used as temporary storage, allowing each summation to be carried out only once per vertex per update.

5   Optimization

It is not difficult to see that the complexity of both Algorithms 1 and 2 is $O(n^2)$, where n is the number of vertices in a digraph. When edges are inserted into or deleted from digraph, there are two cases:

Case 1 the updated edges are incident to a common vertex. In this case, both the path matrix Num and transitive closure matrix M_G^* are updated only one time.

Case 2 the updated edges are not incident to a common vertex. In this case, both the path matrix Num and transitive closure matrix M_G^* will be updated several times.

For the case 2, the procedure of updating matrix can be optimised through caching the records of edge inserting or deletion. That is, we can use several

---

Algorithm 1. Inserting-Edges

1: **Input:** E_v, G_O
2: **Output:** $M_{G_O}^*$ //
3: **for** each $i \in V$ **do**
4:    $From(i) \leftarrow \sum\limits_{(u,v) \in E_v} \#(u,v) \times Num(i,u)$
5: **end for**
6: **for** each $j \in V$ **do**
7:    $To(j) \leftarrow \sum\limits_{(v,u) \in E_v} \#(v,u) \times Num(u,j)$
8: **end for**
9: **for all** $i,j \in V \setminus v$, From(i)$\neq$0 $\bigvee$ To(i)$\neq$0 **do**
10:    Num(i,j) $\leftarrow$ Num(i,j)+[Num(i,v) $\times$ To(j)+Num(v,j) $\times$ From(i)+From(i) $\times$ To(j)]
11: **end for**
12: $j \leftarrow v$
13: **for** each $i \in V \setminus v$, From(i)$\neq$0 **do**
14:    Num(i,j)$\leftarrow$ Num(i,j)+From(i)
15: **end for**
16: $i \leftarrow v$
17: **for** each $j \in V \setminus v$, To(j)$\neq$0 **do**
18:    Num(i,j)$\leftarrow$ Num(i,j)+To(j)
19: **end for**
20: **for all** $i,j \in V$, Num(i,j) has increased from 0 **do**
21:    $M_{G_O}^* \leftarrow 1$
22: **end for**
23: **return** M_G^*

---

arrays *From* (*To*) to record the affected paths. The path matrix *Num* and transitive closure matrix M_G^* is updated only when the number of updated edges is up to a given value.

Furthermore, the strategy of multiple threads can be used to optimize the procedures of identifying affected paths, updating path matrix, and transitive closure matrix.

6   Implementation and Evaluation

In this section, we implement our approach in IncR. We perform experiments to compare IncR with QUONTO [18] and Pellet [13]—QUONT supports the graph-based approach in [21] and Pellet supports module-based incremental classification [13]. We also implement the optimisation in Sect. 5 and perform experiments to test its effectiveness.

We select four widely-used ontologies from Bioportal[1], of which the EL-Galen falls out of OWL 2 QL and need to be approximated by a QL ontology.

---

[1] http://bioportal.bioontology.org/.

Algorithm 2. Deleting-Edges

1: **Input:** E_v, G_O
2: **Output:** $M^*_{G_O}$ //
3: **for** each $i \in V$ and $Num(i, u) \neq 0$ **do**
4:     $From(i) \leftarrow \sum\limits_{(u,v) \in E_v} \#(u, v) * Num(i, u)$
5: **end for**
6: **for** each $j \in V$ and $Num(u, j) \neq 0$ **do**
7:     $To(j) \leftarrow \sum\limits_{(v,u) \in E_v} \#(v, u) * Num(u, j)$
8: **end for**
9: $j \leftarrow v$
10: **for** each $i \in V \setminus v$, From(i)$\neq$0 **do**
11:     Num(i,j)$\leftarrow$ Num(i,j)-From(i)
12: **end for**
13: $i \leftarrow v$
14: **for** each $j \in V \setminus v$, To(j)$\neq$0 **do**
15:     Num(i,j)$\leftarrow$ Num(i,j)-To(j)
16: **end for**
17: **for all** $i, j \in V \setminus v$, From(i)$\neq$0 $\bigvee$ To(i)$\neq$0 **do**
18:     Num(i,j)$\leftarrow$ Num(i,j)-[Num(i,v) $\times$ To(j)+Num(v,j) $\times$ From(i) +From(i) $\times$ To(j)]
19: **end for**
20: **for all** $i, j \in V$, Num(i,j) has decreased to 0 **do**
21:     $M^*_{G_O} \leftarrow 0$
22: **end for**
23: **Return** M^*_G

Table 1 provides basic information: the DL language, the number of atomic concept(**A.C**) and atomic role (**A.R**), the size of ontology (**Size**). The sixth column (**QL Appr.**) presents the size of the OWL 2 QL-approximated version. The last column (**QUONT**) presents the time in seconds taken by QUONTO.

In order to compare IncR with Pellet, we use similar experimental methodology in [13]: for various values of n,

(1) remove n random axioms;
(2) classify the resulting ontology using IncR and Pellet; then, we repeat the following steps 30 times;
(3) randomly remove an additional n axioms and then add back the previously removed n axioms, and further reclassify the ontology using IncR and Pellet.

All the results are gathered during step (3) of the experiment. All tests are performed on a server with Intel Xeon E5620 2.40 GHz CPU, running Windows Server 2008 R2 Enterprise and Java 1.7 with 16 GB of RAM available to JVM.

Table 2 summarizes the results of our experiments for $n = 1, 2, 4, 8$, where Av means average value of the size of affected modules (affected subgraph) and

Table 1. Test ontologies

Ontologies	DL	A.C.	A.R.	Size	QL Appr.	QUONT
Mouse	ALE	2753	1	3463	3463	0.174
Gene	SH	26225	4	42655	42655	1.862
FMA-OBO	ALE	75139	2	119558	119558	5.434
EL-Galen	$ALCH(D)$	23136	950	46457	48026	2.582
Galen	$ALEHIF+$	23141	950	47407	49926	5.735
FMA 1.4	$ALCOIF$	6488	165	18612	18663	2.064
FMA 2.0	$ALCOIF(D)$	41648	148	123610	118181	5.472

Table 2. IncR VS. Pellet.Time in seconds

Ontology	n (updated axioms)	Modules-based approach		Graph-based approach	
		(affected modules) Av/Mx	(updating time) Av/Mx	(affected subgraph) Av/Mx	(updating time) Av/Mx
Mouse	1	96/221	0.021/0.039	44/143	0.002/0.032
Mouse	2	504/812	0.044/0.063	138/218	0.003/0.043
Mouse	4	625/964	0.068/0.072	274/503	0.006/0.061
Mouse	8	852/1811	0.082/0.095	473/96	0.013/0.084
Gene	1	103/571	0.352/1.217	64/187	0.029/0.191
Gene	2	158/1014	0.424/1.422	89/448	0.17/0.464
Gene	4	376/1364	1.63/2.118	204/592	0.35/0.648
Gene	8	849/3294	1.942/2.736	504/916	0.67/1.042
FMA-OBO	1	245/484	1.471/3.742	87/164	0.124/0.67
FMA-OBO	2	369/727	2.921/3.872	148/373	0.723/1.534
FMA-OBO	4	436/1759	3.472/6.532	252/512	0.752/2.463
FMA-OBO	8	1428/3824	4.612/8.344	563/848	1.634/3.942
EL-Galen	1	213/871	0.932/1.684	118/219	0.382/1.647
EL-Galen	2	486/1004	1.364/2.841	164/413	0.431/1.781
EL-Galen	4	869/1864	2.763/4.103	217/593	1.172/2.073
EL-Galen	8	1027/2114	3.142/6.724	572/819	1.038/2. 216
Galen	1	247/921	0.937/2.942	178/221	0.138/1.262
Galen	2	526/1274	1.863/3.814	218/398	0.351/1.684
Galen	4	1069/2564	2.173/5.144	287/643	0.572/1.807
Galen	8	1387/3784	2.36/6.524	598/1019	0.834/2.149
FMA 1.4	1	64/159	0.215/0.746	27/81	0.042/0.415
FMA 1.4	2	129/247	0.368/0.817	46/113	0.061/0.325
FMA 1.4	4	264/635	0.923/1.053	86/212	0.192/0.946
FMA 1.4	8	583/1794	1.032/1.374	163/395	0.312/1.024
FMA 2.0	1	138/297	0.835/1.274	58/171	0.126/0.273
FMA 2.0	2	359/673	0.952/2.638	82/273	0.227/0.952
FMA 2.0	4	496/1469	1.063/3.953	183/451	0.457/1.168
FMA 2.0	8	883/2374	1.61/5.94	383/748	0.636/1.324

Table 3. Optimization using multiple threads

Ontology	n (updated axioms)	One thread	Two threads	Three threads
Mouse	1	0.002/0.032	0.002/0.024	0.002/0.012
Mouse	2	0.003/0.043	0.002/0.022	0.002/0.016
Mouse	4	0.006/0.061	0.003/0.057	0.002/0.032
Mouse	8	0.013/0.084	0.006/0.392	0.004/0.257
Gene	1	0.028/0.192	0.013/0.091	0.007/0.054
Gene	2	0.172/0.468	0.091/0.261	0.048/0.138
Gene	4	0.354/0.643	0.185/0.348	0.097/0.201
Gene	8	0.672/1.047	0.347/0.642	0.182/0.427
FMA-OBO	1	0.124/0.672	0.066/0.384	0.038/0.198
FMA-OBO	2	0.723/1.534	0.382/0.834	0.202/0.506
FMA-OBO	4	0.752/2.463	0.395/1.423	0.205/0.873
FMA-OBO	8	1.634/3.942	0.867/2.062	0.452/1.042
EL-Galen	1	0.382/1.643	0.198/0.884	0.103/0.491
EL-Galen	2	0.431/1.781	0.218/0.901	0.131/0.531
EL-Galen	4	1.172/2.073	0.623/1.103	0.352/0.731
EL-Galen	8	1.038/2. 216	0.536/1. 262	0.278/1. 268
Galen	1	0.138/1.262	0.072/0.694	0.038/0.428
Galen	2	0.351/1.684	0.181/0.901	0.103/0.601
Galen	4	0.572/1.807	0.294/1.007	0.162/0.727
Galen	8	0.834/2.149	0.472/1.297	0.263/0.792
FMA 1.4	1	0.042/0.415	0.028/0.285	0.015/0.157
FMA 1.4	2	0.061/0.325	0.036/0.198	0.019/0.124
FMA 1.4	4	0.192/0.946	0.102/0.516	0.058/0.306
FMA 1.4	8	0.312/1.024	0.164/0.644	0.092/0.413
FMA 2.0	1	0.124/0.275	0.068/0.163	0.039/0.102
FMA 2.0	2	0.237/0.971	0.126/0.572	0.067/0.323
FMA 2.0	4	0.463/1.172	0.257/0.684	0.137/0.428
FMA2.0	8	0.629/1.351	0.366/0.832	0.193/0.624

updating time, and Mx represents maximum value of the size of affected modules (affected subgraph) and updating time in the module-based approach and graph-based approach, respectively. In the column, the time value includes time that taken to classify the ontology by Pellet and time spent in identifying affected modules. In the column 6, the time value includes time spent in identifying affected paths and time in re-computing the transitive closure of all the affected paths.

To evaluate the effectiveness of the optimisation proposed in Sect. 5, we perform experiments by exploiting multiple threads. Table 3 summarizes the results of our experiments.

7   Discussions

We have proposed an graph-based approach to incremental classification in OWL 2 QL ontology by exploiting digraph algorithm, which allows us to efficiently identify the small affected paths and to reuse previous computation. We demonstrate the performance gain indicated by the significant reduction of classification time. Moreover, the graph-based approach make it is easy to implement multiple threads so as to exploit many cores/CPUs available in modern systems. Recently, Zhou and Qi et al. [25, 26] propose a platform-independent approach (called GEL) for parallel reasoning with OWL EL ontologies using graph representation. In the future work, we would apply our incremental strategy to improve the efficiency of GEL.

Acknowledgments. This work is supported by the National Key R&Development Program of China (2016YFB1000603), the National Natural Science Foundation of China (61672377), and the Key Technology Research and Development Program of Tianjin (16YFZCGX00210).

References

1. Motik, B., Patel-Schneider, P.F., Cuenca Grau, B.: OWL 2 Web Ontology Language Direct Semantics, W3C Recommendation (2009)
2. Sidhu, A., Dillon, T., Chang, E., Sidhu, B.S.: Protein ontology development using OWL. In: Proceedings of OWLED 2005 (2005)
3. Golbreich, C., Zhang, S., Bodenreider, O.: The foundational model of anatomy in OWL: experience and perspectives. J. Web Semant. **4**(3), 181–195 (2006)
4. Haarslev, V., Möller, R.: RACER system description. In: Goré, R., Leitsch, A., Nipkow, T. (eds.) IJCAR 2001. LNCS, vol. 2083, pp. 701–705. Springer, Heidelberg (2001). https://doi.org/10.1007/3-540-45744-5_59
5. Sirin, E., Parsia, B., Cuenca Grau, B., Kalyanpur, A., Katz, Y.: Pellet: a practical OWL DL reasoner. J. Web Semant. **5**(2), 51–53 (2007)
6. Tsarkov, D., Horrocks, I.: FaCT++ description logic reasoner: system description. In: Furbach, U., Shankar, N. (eds.) IJCAR 2006. LNCS (LNAI), vol. 4130, pp. 292–297. Springer, Heidelberg (2006). https://doi.org/10.1007/11814771_26
7. Glimm, B., Horrocks, I., Motik, B., Shearer, R., Stoilos, G.: A novel approach to ontology classification. J. Web Semant. **14**(1), 84–101 (2011)
8. Rogers, J., Rector, A.: The GALEN ontology. In: Proceedings of MIE 1996, pp. 174–178 (1996)
9. Golbreich, C., Zhang, S., Bodenreider, O.: The foundational model of anatomy in OWL: experience and perspectives. J. Web Semant. **4**(3), 181–195 (2006)
10. http://www.w3.org/TR/owl2-profiles/

11. Mendez, J., Suntisrivaraporn, B.: Reintroducing CEL as an OWL 2 EL Reasoner. In: Proceedings of the International Workshop on Description Logics (DL 2009), vol. 477 (2009)
12. Kazakov, Y.: Yevgeny, Markus Krötzsch, and František Simančk. The incredible ELK. J. Autom. Reason. **53**(1), 1–61 (2014)
13. Cuenca Grau, B., Halaschek-Wiener, C., Kazakov, Y., Suntisrivaraporn, B.: Incremental classification of description logics ontologies. J. Automated Reason. **44**(4), 337–369 (2010)
14. Suntisrivaraporn, B.: Module extraction and incremental classification: a pragmatic approach for ontologies. In: Proceedings of ESWC 2008, pp. 230–244 (2008)
15. Kazakov, Y., Klinov, P.: Incremental reasoning in OWL EL without bookkeeping. In: Alani, H., Kagal, L., Fokoue, A., Groth, P., Biemann, C., Parreira, J.X., Aroyo, L., Noy, N., Welty, C., Janowicz, K. (eds.) ISWC 2013. LNCS, vol. 8218, pp. 232–247. Springer, Heidelberg (2013). https://doi.org/10.1007/978-3-642-41335-3_15
16. Gupta, A., Mumick, I.S., Subrahmanian, V.S.: Maintaining views incrementally. In: Proceedings of SIGMOD 1993, pp. 157–166 (1993)
17. Motik, B., Nenov, Y., Piro, R., Horrocks, I.: Incremental update of datalog materialisation: the backward/forward algorithm. In: Proceedings of AAAI 2015, pp. 1560–1568 (2015)
18. Calvanese, D., Giacomo, G.D., Lembo, D., Lenzerini, M., Poggi, A., Rodriguez-Muro, M., Rosati, R., Ruzzi, M., Fabio, D.: Savo. The MASTRO system for ontology-based data access. Semantic Web **2**(1), 43–53 (2011)
19. Botoeva, E., Calvanese, D., Santarelli, V., Fabio Savo, D., Solimando, A., Xiao, G.: Beyond OWL 2 QL in OBDA: Rewritings and approximations. In: Proceedings of the 30th AAAI Conference on Artificial Intelligence (AAAI) (2016)
20. Rodriguez-Muro, M., Kontchakov, R., Zakharyaschev, M.: Ontologybased data access: ontop of databases. In: Proceedings of ISWC 2013, pp. 558–573 (2013)
21. Domenico, L., Santarelli, V., Savo, D.F.: A graph-based approach for classifying OWL 2 QL ontologies. In: Proceedings of DL 2013, pp. 747–759 (2013)
22. Baader, F., Calvanese, D., McGuinness, D., Nardi, D., Patel-Schneider, P.F.: The description logic handbook -theory, implementation and applications. Ciudad Y Territorio Estudios Territoriales **62**(7), 497–506 (2001)
23. Calvanese, D., De Giacomo, G., Lembo, D., Lenzerini, M., Rosati, R.: Tractable reasoning and efficient query answering in description logics: the DL-Lite family. J. Automated Reason. **39**(3), 385–429 (2007)
24. King, V., Sagert, G.: A fully dynamic algorithm for maintaining the transitive closure. In: Proceedings of STOC 1999, pp. 492–498 (1999)
25. Zhou, Z., Qi, G., Wu, Z., Shi, J.: A platform-independent approach for parallel reasoning with OWL EL ontologies using graph representation. In: Proceedings of ICTAI 2015, pp. 80–87 (2015)
26. Zhou, Z., Qi, G.: GEL: a platform-independent reasoner for parallel classification with OWL EL ontologies using graph representation. Int. J. Artif. Intell. Tools (IJAIT) **26**(1), 1–33 (2017)

Tensor-Based Representation and Reasoning of Horn-$\mathcal{SHOIQ}$ Ontologies

Zhangquan Zhou[✉]

Southeast University, Nanjing, China
quanzz@seu.edu.cn

Abstract. Recently, tenors have been widely used to encode triples in an RDF graph, which is sometimes called a knowledge graph, for the purpose of knowledge completion and embedding. An interesting question is, can we use tensors to represent OWL ontologies, and handle logical reasoning with ontologies by tensor operations. In this paper, we take the first effort to theoretically build the connection between tensor-based representation and a Horn fragment of OWL, Horn-$\mathcal{SHOIQ}$, i.e., to study how to encode Horn-$\mathcal{SHOIQ}$ ontologies to tensors, and further consider using tensor operations to handle ontology materialization, which is an important logical reasoning service for ontology-based applications. We show that the soundness and completeness of ontology materialization can be guaranteed by using tensor operations.

Keywords: Tensor · Ontology · Horn-$\mathcal{SHOIQ}$ · Materialization

1 Introduction

Tensor has been used as a data structure in many fields for its capability of modeling different problems, and the feasibility of massive processing [7]; many applications are also built on tensor-based platforms, such as TensorFlow,[1] which is published by Google for efficiently processing large tensors. Recently, tenors have also been widely used to encode triples in an RDF graph, which is sometimes called a knowledge graph, for the purpose of knowledge completion [2,3,12,13]. An interesting question is, can we use tensor to equivalently represent the Web Ontology Language[2] (OWL), which is a significant extension of RDF.

There exists work on representing expressive logics by using tensors. One line of the work is to transform logical reasoning to a statistic problem. In the work of [5,6,9,13], a logic rule is encoded to an objective function over tensors, or a constraint condition in an objective function. Logical reasoning is then transformed to a maximization problem. One limit of the work in this line is that the given encoding methods work on facts or logic statements that are in the form of triples describing binary relations (e.g.,⟨*Jack hasMum Helen*⟩), and can

---

[1] https://www.tensorflow.org/.
[2] https://www.w3.org/OWL/.

© Springer Nature Singapore Pte Ltd. 2017
J. Li et al. (Eds.): CCKS 2017, CCIS 784, pp. 30–36, 2017.
https://doi.org/10.1007/978-981-10-7359-5_4

hardly be adapted to logic languages with complex syntaxes, like OWL. Another line of the work focuses on using tensor products (i.e., *Kronecker products* and *inner products*) to represent logic languages and handle logical reasoning. We refer the readers to the two main related work [8,11]. This work studies the reasoning tasks that are provided for special purposes (i.e., query answering on real corpus), but does not aim to study a specific logic language that has formal definitions. Further, there is also no theoretical proofs of the soundness and completeness of reasoning based on tensor operations.

In this paper, we take the first effort to study the problem of using tensors to represent Horn-$\mathcal{SHOIQ}$, an important fragment of OWL, and further consider using tensor operations to handle ontology materialization, which is an important reasoning service in many ontology-based applications. The existing tensor-based methods can hardly be adapted to Horn-$\mathcal{SHOIQ}$, since complex relations (e.g., *concept conjunction* and *existence restriction*) are allowed in Horn-$\mathcal{SHOIQ}$; the task of ontology materialization is also more complicated than the basic reasoning tasks studied in existing works. To address these problems, we propose a new kind of encoding method that can equivalently represent logic statements (i.e., *axioms* and *assertions*) occurring in an Horn-$\mathcal{SHOIQ}$ ontology. Based on this encoding method, we further identify a group of tensor operations to handle materialization. The full version of this work can be found at this address[3].

2 Ontology and Tensor Operations

In this section, we introduce some notions that are used in this paper.

Ontology and Materialization. A Horn-$\mathcal{SHOIQ}$ ontology $\mathcal{O}$ can be defined by a set of axioms of the forms (A1-A11, see the second column in Table 1), *concept assertions* of the form $A(a)$, *role assertions* of the form $R(a, b)$ and *equivalent assertions* of the form $a \approx b$. For more detailed information, we refer the reader to [1]. Given a Horn-$\mathcal{SHOIQ}$ ontology $\mathcal{O}$, we use $\mathtt{mat}(\mathcal{O})$ to denote the result of ontology materialization with respect to $\mathcal{O}$. We assume that $\mathcal{O}$ satisfies *the tractability condition* [4], such that, $\mathtt{mat}(\mathcal{O})$ can be defined by the complete closure of applying the rules in Table 1 (see the third column). Without introducing confusions, we use Rule (Ai) ($i \in \{1, 2, ..., 11\}$) to denote the rule that corresponds to the axiom (Ai). We take Rule (A1) as an example: $A \sqsubseteq B$ and $A(a)$ are called *premises*, $B(a)$ is called a *consequence*, and the common logic element among premises (A in this rule) is called a *joint*. The other rules can be explained similarly. Further, we say that a joint is an *invisible joint* if it does not occur in consequences (like A in the previous example); otherwise it is called a *visible joint* (like c in Rule (A6)).

Tensor Operations [7]. An *order-n tensor* $\mathbf{T}^{(n)}$ can be viewed as an n-dimensional array of real numbers, where each element is written as $\mathbf{T}_{\gamma_1 ... \gamma_n}$ or $[\mathbf{T}]_{\gamma_1 : \gamma_n}$ ($\gamma_k \in \{1, 2, ..., d\}$ for all $k \in \{1, ..., n\}$). We apply three basic tensor

---

[3] https://github.com/Anonymous-account-github/CCKS17/.

Table 1. Ontology axioms, rules for materialization and corresponding tensor operations

	Axioms	Rules	Tensor Operations
(A1)	$A \sqsubseteq B$	$A \sqsubseteq B, A(a) \vdash B(a)$	$[\phi_{A1}(\mathfrak{O})]_{x_1:x_7} = \operatorname{T}^{\mathrm{ca}}_{x_1}(\mathfrak{O}_{y_1 y_2 y_3 x_2 x_3 y_6 x_6} \operatorname{T}^{A1}_{y_1})$ $(\mathfrak{O}_{z_1 y_2 y_3 x_4 x_5 y_6 x_7} \operatorname{T}^{\mathrm{ca}}_{z_1})$
(A2)	$A_1 \sqcap A_2 \sqsubseteq B$	$A_1 \sqcap A_2 \sqsubseteq B, A_1(a), A_2(a) \vdash B(a)$	$[\phi^{A2}(\mathfrak{O})]_{x_1:x_7} = \operatorname{T}^{\mathrm{ca}}_{x_1}(\mathfrak{O}_{y_1 y_2 y_3 y_4 y_5 x_2 x_3} \operatorname{T}^{A2}_{y_1})$ $(\mathfrak{O}_{z_1 y_2 y_3 x_4 x_4 x_6 x_6} \operatorname{T}^{\mathrm{ca}}_{z_1})$ $(\mathfrak{O}_{k_1 y_4 y_5 x_4 x_5 x_6 x_7} \operatorname{T}^{\mathrm{ca}}_{k_1})$
(A3)	$A \sqsubseteq \forall R.B$	$A \sqsubseteq \forall R.B, A(a), R(a,b) \vdash B(b)$	$[\phi^{A3}(\mathfrak{O})]_{x_1:x_7} = \operatorname{T}^{\mathrm{ca}}_{x_1}(\mathfrak{O}_{y_1 y_2 y_3 y_4 y_5 x_2 x_3} \operatorname{T}^{A3}_{y_1})$ $(\mathfrak{O}_{z_1 y_2 y_3 x_4 x_5 x_6 x_7} \operatorname{T}^{\mathrm{ca}}_{z_1})$ $(\mathfrak{O}_{k_1 y_4 y_5 x_4 x_5 x_4 x_5} \operatorname{T}^{\mathrm{ra}}_{k_1})$
(A4)	$\exists R.A \sqsubseteq B$	$\exists R.A \sqsubseteq B, R(a,b), A(b) \vdash B(a)$	$[\phi^{A4}(\mathfrak{O})]_{x_1:x_7} = \operatorname{T}^{\mathrm{ca}}_{x_1}(\mathfrak{O}_{y_1 y_2 y_3 y_4 y_5 x_2 x_3} \operatorname{T}^{A4}_{y_1})$ $(\mathfrak{O}_{z_1 y_2 y_3 x_4 x_5 x_6 x_7} \operatorname{T}^{\mathrm{ca}}_{z_1})$ $(\mathfrak{O}_{k_1 y_4 y_5 x_4 x_5 x_4 x_5} \operatorname{T}^{\mathrm{ra}}_{k_1})$
(A5)	$A \sqsubseteq \exists R.B$	$A \sqsubseteq \exists R.B, A(a) \vdash R(a, o^A_{R,B}), B(o^A_{R,B})$	$[\phi^{A5}(\mathfrak{O})]_{x_1:x_7} = \operatorname{T}^{\mathrm{ra}}_{x_1}(\mathfrak{O}_{y_1 y_2 y_3 x_2 x_3 x_6 x_7} \operatorname{T}^{A5.1}_{y_1})$ $(\mathfrak{O}_{z_1 y_2 y_3 x_4 x_5 x_6 x_6} \operatorname{T}^{\mathrm{ca}}_{z_1})$ $+\operatorname{T}^{\mathrm{ca}}_{x_1}(\mathfrak{O}_{l_1 k_2 k_3 k_2 x_3 x_4 x_5} \operatorname{T}^{A5.2}_{k_1})$ $(\mathfrak{O}_{l_1 k_2 k_3 l_4 l_4 x_6 x_7} \operatorname{T}^{\mathrm{ca}}_{l_1})$
(A6)	$A \sqsubseteq^{\leq} 1R.B$	$A \sqsubseteq^{\leq} 1R.B, A(a), R(a,b), R(a,c)$ $, B(c) \vdash b \approx c$	$[\phi^{A6}(\mathfrak{O})]_{x_1:x_7} = \operatorname{T}^{\mathrm{eq}}_{x_1}(\mathfrak{O}_{y_1 y_2 y_3 y_4 y_5 y_6 y_7} \operatorname{T}^{A6}_{y_1})$ $(\mathfrak{O}_{z_1 y_2 y_3 x_4 x_5 x_6 x_6} \operatorname{T}^{\mathrm{ca}}_{z_1})$ $(\mathfrak{O}_{k_1 k_2 y_4 k_4 x_4 x_2 x_3} \operatorname{T}^{\mathrm{ra}}_{k_1})$ $(\mathfrak{O}_{l_1 k_2 y_5 k_4 x_5 l_6 x_4} \operatorname{T}^{\mathrm{ra}}_{l_1})$ $(\mathfrak{O}_{m_1 y_6 y_7 l_6 x_5 x_6 x_7} \operatorname{T}^{\mathrm{ca}}_{m_1})$
(A7)	$A \sqsubseteq \neg B$	$A \sqsubseteq \neg B, A(a) \vdash \neg B(a)$	$[\phi^{A7}(\mathfrak{O})]_{x_1:x_7} = \operatorname{T}^{\mathrm{ng}}_{x_1}(\mathfrak{O}_{y_1 y_2 y_3 x_2 x_3 x_6 x_7} \operatorname{T}^{A7}_{y_1})$ $(\mathfrak{O}_{z_1 y_2 y_3 x_4 x_5 x_6 x_7} \operatorname{T}^{\mathrm{ca}}_{z_1})$
(A8)	$A \sqsubseteq \{a\}$	$A \sqsubseteq \{a\}, A(b) \vdash b \approx a$	$[\phi^{A8}(\mathfrak{O})]_{x_1:x_7} = \operatorname{T}^{\mathrm{ng}}_{x_1}(\mathfrak{O}_{y_1 y_2 y_3 x_4 x_5 x_6 x_7} \operatorname{T}^{A8}_{y_1})$ $(\mathfrak{O}_{z_1 y_2 y_3 x_2 x_3 x_6 x_6} \operatorname{T}^{\mathrm{ca}}_{z_1})$
(A9)	$R \sqsubseteq S$	$R \sqsubseteq S, R(a,b) \vdash S(a,b)$	$[\phi^{A9}(\mathfrak{O})]_{x_1:x_7} = \operatorname{T}^{\mathrm{ra}}_{x_1}(\mathfrak{O}_{y_1 y_2 y_3 x_2 x_3 y_6 y_6} \operatorname{T}^{A9}_{y_1})$ $(\mathfrak{O}_{z_1 y_2 y_3 x_4 x_5 x_6 x_7} \operatorname{T}^{\mathrm{ra}}_{z_1})$
(A10)	$R \sqsubseteq S^-$	$R \sqsubseteq S^-, R(a,b) \vdash S(b,a)$	$[\phi^{A10}(\mathfrak{O})]_{x_1:x_7} = \operatorname{T}^{\mathrm{ra}}_{x_1}(\mathfrak{O}_{y_1 y_2 y_3 x_2 x_3 y_6 y_6} \operatorname{T}^{A10}_{y_1})$ $(\mathfrak{O}_{z_1 y_2 y_3 x_6 x_7 x_4 x_5} \operatorname{T}^{\mathrm{ra}}_{z_1})$
(A11)	$Tra(R)$	$Tra(R), R(a,b), R(b,c) \vdash R(a,c)$	$[\phi^{A11}(\mathfrak{O})]_{x_1:x_7} = \operatorname{T}^{\mathrm{ra}}_{x_1}(\mathfrak{O}_{y_1 y_2 y_3 y_4 y_5 y_5} \operatorname{T}^{A11}_{y_1})$ $(\mathfrak{O}_{z_1 x_2 y_2 x_4 x_5 x_6 x_7} \operatorname{T}^{\mathrm{ra}}_{z_1})$ $(\mathfrak{O}_{k_1 x_2 x_3 x_6 x_7 x_6 x_7} \operatorname{T}^{\mathrm{ra}}_{k_1})$

operations in this paper, *tensor addition, Kronecker product* and *inner product*, which are introduced below. Tensor addition is denoted by the operator $+$, and used as $\mathbf{U}^{(n)} + \mathbf{V}^{(n)}$, where $\mathbf{U}^{(n)}$ and $\mathbf{V}^{(n)}$ are two addends of $\mathbf{T}^{(n)}$. Kronecker product, denoted by $\otimes$, and can be performed among any two or more tensors like $\mathbf{U}^{(n)} \otimes \mathbf{V}^{(m)} = \mathbf{T}^{(n+m)}$ where $\mathbf{T}_{\gamma_1 \dots \gamma_n \gamma'_1 \dots \gamma'_m} = \mathbf{U}_{\gamma_1 \dots \gamma_n} \mathbf{V}_{\gamma'_1 \dots \gamma'_m}$. For a tensor $\mathbf{T} = \mathbf{U}_1 \otimes \dots \otimes \mathbf{U}_k$, we say that $\mathbf{U}_i$ is the ith *Kronecker term* of $\mathbf{T}$ where $i \in \{1, \dots, k\}$. Inner product, denoted by $\bullet$, is defined as follows: $\mathbf{U}^{(n)} \bullet_{(i,j)} \mathbf{V}^{(m)} = \mathbf{T}^{(n+m-2)}$, where i (resp., j) points to the ith (resp., j^{th}) order of $\mathbf{U}^{(n)}$ (resp., $\mathbf{V}^{(m)}$) for $i \in \{1, \dots, n\}$ and $j \in \{1, \dots, m\}$. Each element in the tensor $\mathbf{T}^{(n+m-2)}$ is written as $\mathbf{T}_{\gamma_1 \dots \gamma_{i-1} \gamma_{i+1} \dots \gamma_n \gamma'_1 \dots \gamma'_{j-1} \gamma'_{j+1} \dots \gamma'_m} = \sum_{\beta=1}^{d} \mathbf{U}_{\gamma_1 \dots \gamma_{i-1} \beta \gamma_{i+1} \dots \gamma_n} \mathbf{V}_{\gamma'_1 \dots \gamma'_{j-1} \beta \gamma'_{j+1} \dots \gamma'_m}$; we call the ith order of $\mathbf{U}$ and the jth order of $\mathbf{V}$ a pair of the *inner product orders* (an *IPO pair* for short). For simplicity, we apply *the Einstein Summation Convention* in inner products. Inner products can be easily extended to general operations on two or more IPO pairs.

3   A Tensor-Based Representation for Ontologies

In this section, we discuss how to encode Horn-$\mathcal{SHOIQ}$ ontologies through tensors.

Basic Ideas. In this work, we follow the idea of [10] to encode ontologies based on tensor products. The basic elements (concepts, roles and individuals) in a Horn-$\mathcal{SHOIQ}$ ontology can be intuitively encoded to order-1 tensors (namely vectors), which are also called *basic tensors*. For each rule of (A1)–(A11), the premises can be matched by performing inner products on the IPO pairs that point to joints [6,8,11]. We call these inner products *joint-based products*. Consider Rule (A1). The application of Rule (A1) can be performed by using the inner product $(\mathbf{A} \otimes \mathbf{B}) \bullet_{(1,1)} (\mathbf{A} \otimes \mathbf{a})$, where the IPO pair $(1,1)$ points to the basic tensor $\mathbf{A}$. The correctness of joint-based products is guaranteed by *the orthogonality condition* [11].

Symmetric Order-2 Basic Tensors. One can check that, the above joint-based products can hardly work for the cases where visible joints occur. This is because that a joint-based product would eliminate joints, while visible joints should be preserved. To address this issue, we consider using *symmetric order-2* tensors [7]. A symmetric order-2 tensor can be defined as a self Kronecker product over an order-1 tensor, i.e., $\mathbf{V} \otimes \mathbf{V}$ where $\mathbf{V}$ is an order-1 tensor. It can be checked that $(\mathbf{V} \otimes \mathbf{V}) \bullet_{(n,m)} (\mathbf{V} \otimes \mathbf{V}) = (\mathbf{V} \otimes \mathbf{V})$ where $n, m \in \{1,2\}$. In this way, when applying Rule (A2) by tensor operations, if the joint $\mathbf{a}$ is a symmetric order-2 tensor, $\mathbf{a}$ can be preserved after performing a joint-based product between $\mathbf{A}_1 \otimes \mathbf{a}$ and $\mathbf{A}_2 \otimes \mathbf{a}$. In this work, we use symmetric order-2 tensors as basic tensors.

Encoding Axioms and Assertions. Based on the above idea, we can encode an ontology by performing a tensor addition over all tensorial axioms and assertions. Note that, all addends in an addition should have the same order. To this end, we give a transformation function τ over all axioms and assertions in a given ontology $\mathcal{O}$ (see Table 2).

We use the unique tensors $\mathbf{T}^{\varphi}$ ($\varphi \in \{\text{A1}, ..., \text{A11}\}$), $\mathbf{T}^{\text{ca}}$, $\mathbf{T}^{\text{ra}}$ and $\mathbf{T}^{\text{eq}}$ for indicating the type of an axiom or assertion. For axioms or assertions of the forms $A \sqsubseteq B$, $A \sqsubseteq \neg B$, $A \sqsubseteq \{a\}$, $R \sqsubseteq S$, $R \sqsubseteq S^-$, $\text{Tra}(R)$, $A(a)$ and $a \approx b$, they can be transformed to tensors of lower orders compared to the rest ones, since we can use less than three basic tensors to encode these axioms or assertions. We introduce a new symmetric order-2 tensor $\theta = \theta \otimes \theta$, to make the final tensorial axioms and assertions be at the same order (see the Eqs. (1, 7–12, 14)). Based on the above encoding method, a given ontology $\mathcal{O}$ can be transformed to an order-7 tensor which is the summation of all tensorial axioms and assertions.

4   Materialization via Tensor Operations

Based on the above encoding method, we discuss in this section how to handle materialization by tensor operations, i.e., how to handle the rules in Table 1 by joint-based products.

Table 2. The transformation function τ

$\tau(A \sqsubseteq B)$	=	$T^{A1} \otimes A \otimes B \otimes \theta$	(1)
$\tau(A_1 \sqcap A_2 \sqsubseteq B)$	=	$T^{A2} \otimes A_1 \otimes A_2 \otimes B$	(2)
$\tau(A \sqsubseteq \forall R.B)$	=	$T^{A3} \otimes A \otimes R \otimes B$	(3)
$\tau(\exists R.A \sqsubseteq B)$	=	$T^{A4} \otimes R \otimes A \otimes B$	(4)
$\tau(A \sqsubseteq \exists R.B)$	=	$T^{A5} \otimes A \otimes R \otimes B$	(5)
$\tau(A \sqsubseteq\leq 1R.B)$	=	$T^{A6} \otimes A \otimes R \otimes B$	(6)
$\tau(A \sqsubseteq \neg B)$	=	$T^{A7} \otimes A \otimes B \otimes \theta$	(7)
$\tau(A \sqsubseteq \{a\})$	=	$T^{A8} \otimes A \otimes a \otimes \theta$	(8)
$\tau(R \sqsubseteq S)$	=	$T^{A9} \otimes R \otimes S \otimes \theta$	(9)
$\tau(R \sqsubseteq S^-)$	=	$T^{A10} \otimes R \otimes S \otimes \theta$	(10)
$\tau(\mathtt{Tra}(R))$	=	$T^{A11} \otimes R \otimes \theta \otimes \theta$	(11)
$\tau(A(a))$	=	$T^{ca} \otimes A \otimes a \otimes \theta$	(12)
$\tau(R(a,b))$	=	$T^{ra} \otimes R \otimes a \otimes b$	(13)
$\tau(a \approx b)$	=	$T^{eq} \otimes a \otimes b \otimes \theta$	(14)

Decomposition and Combination. The basic idea of performing materialization over tensors is as follows. A tensorial ontology $\mathfrak{O}$ is a summation of all tensorial axioms and assertions. Thus, we consider first *decomposing* $\mathfrak{O}$ to get needed tensors, then, performing joint-based products over decomposed tensors, and, finally, *combing* result tensors to $\mathfrak{O}$.

We take Rule (A1) as an example. This rule has two premises, i.e., $A \sqsubseteq B$ and $A(a)$. Thus, we can just focus on the tensors with Kronecker term T^{A1} and T^{ca}. We first decompose the given tensorial ontology $\mathfrak{O}$ through $T_1 = \mathfrak{O} \bullet_{(1,1)} T^{A1}$ and $T_2 = \mathfrak{O} \bullet_{(1,1)} T^{ca}$ to get all tensors with T^{A1} and T^{ca} as the first Kronecker term respectively. Then we perform a joint-based product between T_1 and T_2 with respect to the joint. Finally, the result of the previous product is added back to $\mathfrak{O}$. The new tensorial ontology $\mathfrak{O}$ is essentially the result by applying Rule (A1) once. The formulation of the above process is shown in the first row and the last column in Table 1. The tensor-based operations of other rules also shown in Table 1. For lack of space, we do not give detailed explanation of these rules. We refer the readers to the full version of this paper.

Complete Materialization. It should be noted that assertions in $\mathtt{mat}(\mathcal{O})$ are obtained by applying the rules in Table 1 iteratively until no more consequence can be obtained. Thus, we should also apply all the eleven tensor-based operations iteratively to conduct a complete materialization. To this end, we define a new operation $\Phi(\mathfrak{O}) = \Sigma_\varphi \phi^\varphi(\mathfrak{O})$ where $\varphi \in \{A1, ..., A11\}$. Further, we give the following operations: $\Phi^0(\mathfrak{O}) = \mathfrak{O}$, $\Phi^{i+1}(\mathfrak{O}) = \Phi(\Phi^i(\mathfrak{O})) + \Phi^i(\mathfrak{O})$, where i is an integer and $i \geq 1$. The first addend of the operation $\Phi^{i+1}(\mathfrak{O})$ is to apply Φ once over $\Phi^i(\mathfrak{O})$, which can be viewed as an updated tensorial ontology with tensors newly computed from $\Phi^{i-1}(\mathfrak{O})$. To include old tensors, the addend $\Phi^i(\mathfrak{O})$ has to be added to $\Phi^{i+1}(\mathfrak{O})$. We use Theorem 1 to show the soundness and completeness of Φ.

Theorem 1. *Given an ontology $\mathcal{O}$, $\mathfrak{O}$ is the corresponding tensor of $\mathcal{O}$ based on a basic vector set $\mathbb{V}$ and the transformation function τ. The basic vector set $\mathbb{V}$ satisfies the orthogonality condition. For any integer $i \geq 0$ and any assertion α, we have that:*

(Soundness) $\Phi^i(\mathfrak{O}) \bullet \tau(\alpha) \geq 1$ implies $\alpha \in \mathtt{mat}(\mathcal{O})$;[4]

(Completeness) there exists an integer N such that if $\alpha \in \mathtt{mat}(\mathcal{O})$ then $\Phi^N(\mathfrak{O}) \bullet \tau(\alpha) \geq 1$.

5 Conclusions and Future Work

In this paper, we provided a novel method to encode a Horn-$\mathcal{SHOIQ}$ ontology to tensors by using symmetric order-2 tensors. The new encoding method avoids the problems caused by handing visible joints. We further showed the correctness of the encoding method. We then presented a method to perform ontology materialization by using tensor operations. We showed the soundness and completeness of the proposed method. As one future work, we will implement a system for ontology materialization based on high-performance tensor-based platforms. As another future work, we consider incorporating the techniques of knowledge completion with our methods. Specifically, the 11 operations $\phi^{\mathtt{A1}}...\phi^{\mathtt{A11}}$ can be formulated as constraints in a linear programming problem by following the idea given in [13]. This would improve the accuracy for knowledge completion.

References

1. Baader, F., Calvanese, D., McGuinness, D.L., Nardi, D., Patel-Schneider, P.F.: The Description Logic Handbook: Theory, Implementation and Applications. Cambridge University Press, Cambridge (2003)
2. Bordes, A., Usunier, N., García-Durán, A., Weston, J., Yakhnenko, O.: Translating embeddings for modeling multi-relational data. In: Proceedings of NIPS, pp. 2787–2795 (2013)
3. Bordes, A., Weston, J., Collobert, R., Bengio, Y.: Learning structured embeddings of knowledge bases. In: Proceedings of AAAI (2011)
4. Carral, D., Feier, C., Grau, B.C., Hitzler, P., Horrocks, I.: Pushing the boundaries of tractable ontology reasoning. In: Mika, P., et al. (eds.) ISWC 2014. LNCS, vol. 8797, pp. 148–163. Springer, Cham (2014). https://doi.org/10.1007/978-3-319-11915-1_10
5. Chang, K., Yih, W., Yang, B., Meek, C.: Typed tensor decomposition of knowledge bases for relation extraction. In Proceedings of EMNLP, pp. 1568–1579 (2014)
6. Grefenstette, E.: Towards a formal distributional semantics: simulating logical calculi with tensors. CoRR (2013)
7. Kolda, T.G., Bader, B.W.: Tensor Decompositions and Applications. Sandia National Laboratories, Albuquerque (2007)

---

[4] Here, we use '≥ 1' to include the situations where redundant occurs. Specifically, if $\tau(\alpha)$ occurs as an addend in $\Phi^i(\mathfrak{O})$ more than once, it can be checked that $\Phi^i(\mathfrak{O}) \bullet \tau(\alpha) \geq 1$.

8. Lee, M., He, X., Yih, W., Gao, J., Deng, L., Smolensky, P.: Reasoning in vector space: an exploratory study of question answering. CoRR (2015)
9. Rocktäschel, T., Bosnjak, M., Singh, S., Riedel, S.: Low-dimensional embeddings of logic. In: Proceedings of ACL Workshop on Semantic Parsing, pp. 45–49 (2014)
10. Smolensky, P.: Tensor product variable binding and the representation of symbolic structures in connectionist systems. J. Artif. Intell. **46**(1–2), 159–216 (1990)
11. Smolensky, P., Lee, M., He, X., Yih, W., Gao, J., Deng, L.: Basic reasoning with tensor product representations. CoRR (2016)
12. Socher, R., Chen, D., Manning, C.D., Ng, A.Y.: Reasoning with neural tensor networks for knowledge base completion. In: Proceedings of NIPS, pp. 926–934 (2013)
13. Wang, Q., Wang, B., Guo, L.: Knowledge base completion using embeddings and rules. In Proceedings of IJCAI, pp. 1859–1866 (2015)

Attention-Based Event Relevance Model for Stock Price Movement Prediction

Jian Liu[1,2(✉)], Yubo Chen[1], Kang Liu[1,2], and Jun Zhao[1,2]

[1] National Laboratory of Pattern Recognition, Institute of Automation,
Chinese Academy of Sciences, Beijing 100190, China
{jian.liu,yubo.chen,kliu,jzhao}@nlpr.ia.ac.cn
[2] University of Chinese Academy of Sciences, Beijing 100049, China

Abstract. Stock prices, in general, can be affected by world events such as wars, natural disasters, government policies, etc. However, the correlations between events and stock prices are often implicit and the influences of events on stock prices can be in indirect ways and act in chain reactions, which brings essential difficulties for precise market prediction. In this paper, we propose an attention-based event relevance model (ATT-ERNN) to explicitly model event relevance for predicting stock price movement. Specifically, in our model, we use long short-term memory neural network (LSTM) and convolution neural network (CNN) to encode event information and stock information to distributional representations. After that, we employ attention mechanism to find related events for each stock to do price movement prediction. Attention weights in our model have a quantitative interpretation as the relevance degree of events affecting the price of a specific stock. We have conduct extensive experiments on a manually collected real-world dataset. Experimental results show the superiority of our model over many baselines, which proves the effectiveness of our model in this prediction problem.

Keywords: Stock price movement prediction
Attention-based neural network · Event relevance modeling
Deep learning

1 Introduction

Stock market prediction is the act of trying to determine the future value of a company stock [16]. The successful prediction of a stock's future price can yield significant profit obviously, making the prediction problem almost the hottest topic in finance field and artificial intelligence area.

The well-known efficient-market hypothesis (EMH) [9,15] suggests that stock prices reflect all currently available information and any price changes based on the newly revealed relevant information. However, finding relevant information that contribute to the change of price for each individual stock faces essential difficulties because the correlations between daily events and their effect on the stock prices are usually implicit. Besides, the influences of events to stock prices

© Springer Nature Singapore Pte Ltd. 2017
J. Li et al. (Eds.): CCKS 2017, CCIS 784, pp. 37–49, 2017.
https://doi.org/10.1007/978-981-10-7359-5_5

can occur in indirect ways and act in chain reactions, which sets obstacles for precise market prediction.

Consider a real-world example: America was hit by hurricane Sandy in 2012, which was one of the worst storms on record that battered the Eastern United States, causing up to $20 billion in damage. This disaster drove many stocks down, including companies of the insurance industry, transportation industry, etc. However, according to financial reports, some companies benefit from this disaster! The famous Internet-based company Facebook was one of the examples. At the first glance, there seems no correlation between "hurricane causes damage" and "stock price of Facebook goes up", but when taking a deeper inspections, we can find a plausible explanation: During a storm, people will want to check in on family, friends and loved ones, as well as share photos of storm and its impact. People can't go out, so they have to use Facebook, a popular message delivery platform to communicate with other people. That is, the storm makes people gain a higher dependence on Facebook than usual, and then makes a higher expectation of investors on Facebook, causing the stock price to boost.

We argue that, for making precise prediction of a specific stock, the core question is to find which events provide related information can affect the price. Traditional approaches usually depend on complex features [1,4,5,21–23] to capture the correlations, which requires a lot of domain knowledge and takes huge human ingenuity. Different from these approaches, in this paper, we propose an end-to-end attention-based event relevance model (ATT-ERNN) to learn the hidden correlations from data. Our model uses neural network based encoder to encode event information and stock information to unified semantic representations, and leverages attention mechanism to model the correlations between event-stock pairs. For a specific stock, higher attention weights show that the events are more important for affecting the price. We have conduct extensive experiments, and the experimental results show the effectiveness of the proposed model.

We summary our contributions as follows:

(1) We propose a novel end-to-end attention-based event relevance model to do stock price movement prediction. To our best knowledge, this is the first work to introduce attention mechanism to stock price movement prediction.
(2) We conduct extensive experiments on a real-world dataset. We compare our model with several baselines, and experimental results show the superiority of our proposed model.
(3) We also exploit the effect of events count and short-term, medium-term, long-term influence for stock price movement prediction.

The remaining of this paper is organized as this: Sect. 2 reviews the related works; Sect. 3 runs into the architecture of the proposed ATT-ERNN model; Sect. 4 illustrates the dataset construction and experimental results and Sect. 5 concludes the whole paper.

2 Related Works

Over the years, various methods have been proposed for stock price prediction. Traditional approaches include using historical stock prices to predict the current price [1, 21], using user sentiment to predict stock price [4, 5, 22, 23], etc. Indeed, price history and user sentiment provide important clues for forecasting the movement of stock price. However, only use these sole evidence and completely ignoring the finance news makes these approaches behave poorly. Meanwhile, traditional approaches rely heavily on human ingenuity to design features, which often requires expert and domain knowledge.

Recently, with the development of deep learning, many neural network based models start to appear in stock price prediction. Ronnqvist et al. [17] use news document to predict potential bank crisis. In their work, they train a combined neural network to predict whether a sentence describing a bank crisis. Ding et al. [6, 7] extract structured event information from news headlines to predict the S&P 500 index. Their structure representation of events combining with CNN achieves an improvement of 6% over state-of-the-art methods. Feuerriegel et al. [8] publish a recurrent autoencoder model to predict Germany stock profit. Their model outperforms random forests as a benchmark with the accuracy by 5.66%. Deep learning based methods achieve better performance for their automatic feature extraction ability and powerful function fitting ability.

In deep learning community, attention-based neural networks gain huge popularity [3, 18, 25], which have made successful applications in neural machine translation [3], abstractive document summarization [18], image caption generation [25], etc. The advantage of the attention mechanism is that it can find the hidden correlations between two different modalities, and make the model focus only on some particular parts.

Guided by the intuition of attention mechanism, for stock price movement prediction problem, we take advantage of it to find the hidden correlations between events and stock prices, and the experiment results show its effectiveness. We believe that if combined with traditional feature-based methods, our model can get more better performance.

3 Methodology

We present the architecture of ATT-RENN model in this section. The ATT-RENN model can be further separated into three sub models: event encoding model, event relevance computing model, and price movement prediction model. In the following, we illustrate the details of each sub model.

3.1 Problem Formulation

We treat stock price movement prediction problem as a **binary classification problem**. Specifically, a stock s with its meta-information (meta-information includes company description, company location, output products, etc.) is given and the current stock price is p. Meanwhile a set of events represented as $E = \{e_i\}$ happens. The stock price changes to q after a fixed period. We assume the change of price is the consequence of the events. Comparing q with p, we can figure out whether the price goes up or down.

Thus, the problem is simplified as binary classification given s and E. We denote the result as $t \in \{0, 1\}$, which indicates the stock price goes down (t equals 0) or goes up (t equals 1). Note that, for different stocks s_j and s_k, the events set $E = \{e_i\}$ can be same. While the price movements of s_j and s_k can behave completely differently, which calls for exploiting correlations between events and stocks. We make the assumption that the influences of the events will last for a fixed period and are independent of other factors. In our experiments, we set the fixed period as one day, one week, one month to exploit the short-term, medium-term, long-term influence.

3.2   Attention-Based Event Relevance Model

The ATT-RENN model consists of three sub models: event encoding model, event relevance computing model, and prediction model. We summary the function of each sub models as below:

Event encoding model aims to encode event content to unified semantic representation. It utilizes word vectors [14] as basic lexical features and uses long short-term memory neural network (LSTM) as the encoder cell. The end-to-end structure avoids complex feature engineering for representing events.

Event relevance computing model aims to model the relevance of each event e_i with respect to a specific stock s. The attention weights have a interpretation as relevance weights of events attribute to the change of stock price.

Prediction model is a fast-forward neural network with nonlinear hidden layers. The nonlinear structure makes the model have the ability to learn complex interactions between different factors.

The overall architecture of the model is plotted in Fig. 1.

Event Encoding Model. The same world event can be described in different expression ways. This variety makes the representation space for events very sparse. Several approaches [6,7] represent the event as tuple $<S, V, O>$ to gain generalization. However, we argue that this representation is oversimplified and might lose a lot of valuable information. Instead, in our model, we propose an LSTM-based encoder to encode the entire event content to a distributional representation to tackle with sparsity problem.

LSTM belongs to Recurrent Neural Networks (RNN), and it contains three gates to maintain the information passing to capture long-term dependency [11].

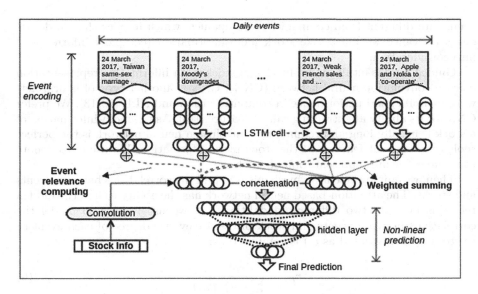

Fig. 1. The overall architecture of the ATT-RENN model

LSTM shows the promising result in sentence encoding in many NLP applications [10, 20]. The computations of LSTM cells are:

$$f_t = \sigma(W_f z_t + U_f h_{t-1} + b_f) \tag{1}$$

$$i_t = \sigma(W_i z_t + U_i h_{t-1} + b_i) \tag{2}$$

$$o_t = \sigma(W_o z_t + U_o h_{t-1} + b_o) \tag{3}$$

$$c_t = f_t \circ c_{t-1} + i_t \circ \sigma(W_c z_t + U_c h_{t-1} + b_c) \tag{4}$$

$$h_t = o_t \circ \sigma(c_t) \tag{5}$$

Where z_t is the input of time t, and h_t is the hidden state of time t. In event encoding model, the content of each event can be represented as $e_i = \{w_{i1}, w_{i2}, ..., w_{in}\}$, where in is the content length of event e_i.

We convert all of the words to word vectors as the inputs to LSTM event encoder, and after encoding, we use the final hidden state as the event representation. Thus we transfer event set $E = \{e_i\}$ to unified semantic representations $X = \{x_i\}$, where x_i is the corresponding representation of e_i.

We prefer LSTM as event encoder for two main considerations: (1) LSTM has the ability to model long sequence and semantic of sentence. (2) The content lengths of events are usually different, and RNN is good at dealing with the variable length problem.

Event Relevance Computing Model. The same events might cause different effects on different stocks. If we want to predict the stock price for a specific company, we must find which events are relevant. Note that, the relevance of events are different from companies to companies, which implies that to decide event relevance we must take stock meta-information (company information) into consideration.

Guided by this intuition, we firstly get stock meta-information representation using a convolution neural network (CNN). CNN is another type of neural network exhibits good performance in computer vision and NLP [12,13]. We prefer CNN for its ability at extracting salient features [13]. As the meta-information of a stock is usually long and redundant, convolution neural network is the perfect tools to get salient features of the stock meta-information. We use c_i to denote the salient feature representation of stock s_i.

Then, we employ attention mechanism to find the correlations between events and stocks. The attention-based neural network has the ability to find the hidden relations between two different modalities, and we use it here to model the correlations between events and stock prices. Relevance degree of each event e_j to stock s_i (represented as c_i) is computed as:

$$r_{i,j} = \frac{exp(p_{i,j})}{\sum_j exp(p_{i,j})} \tag{6}$$

where

$$p_{i,j} = W_r[c_i; x_j] + b_r \tag{7}$$

The comprehensive information from all events is computed as:

$$a_i = \sum_j r_{i,j} x_j \tag{8}$$

We prefer linear attention (Eq. 7) for its good performance in our experiments. In Eq. 9, we use a weighted sum to get comprehensive information from all events according to their relevance weights with respect to stock s_i.

Prediction Model. Before making the final prediction, we concatenate the comprehensive information a_i with the stock meta-information representation c_i. Then, we add a nonlinear hidden layer (with ReLu activation function) to let the model learn complex interactions. Finally, a softmax output layer yields the final prediction probability:

$$H_i = ReLu(W_H[a_i; c_i] + b_H) \tag{9}$$
$$O_i = Softmax(W_O H_i + b_o) \tag{10}$$

H_i is the non-linear transformation, and O_i is a two-value vector indicates the probability of price down or up of stock s_i. O_{i0} represents the probability of "goes down", and O_{i1} represents the probability of "goes up" respectively.

3.3   Model Training

We use the cross-entry loss as the loss function for model optimization. The overall loss can be represented as:

$$J(\Theta) = -\sum_i \sum_j \log(o_{i,j}) + \lambda(\Theta) \tag{11}$$

Where Θ represents all the parameters of our model, $o_{i,j}$ is the corresponding probability output by the model for stock s_i in a specific date j. λ is the regularization weight. We also add dropout layer [19] to prevent over fitting on training phase. We use Adam as the optimization method, which gives the best result among all optimization methods in this dataset.

4   Experiments and Results

4.1   Dataset Construction

Stock price movement prediction has a large practical application value. To make the evaluations and conclusions convincing and close to the real market, we use real-world data to conduct our experiments. The collecting process can be further divided into two parallel branches:

(1) **Event Data.** We obtain news articles from news publishing websites (in our experiments Baidu News[1] and Sina News[2]). We denote each event as $e_i =$ $<tc_i, d_i>$, where tc_i is the text describing an event happens, and d_i is the date associated with the event. We use ICTCLAS[3] to do data preprocessing (word segmentation, stop words removal, etc.). The following shows an event instance (we translate it to English):

   $e_i = $ <China Internet Finance Association was founded, 2016–10-21>

(2) **Stock Price History.** We obtain stock price history from finance related website. Associated with each stock are the price history and the stock meta-information (company description and output products in our experiments). We use the closing price for comparison because in general closing price represents the most up-to-date valuation of a stock.

The statistics of the dataset are summarized in Table 1.

We align stock price history with events according to time annotations. To conduct our experiments, for each industry, we sample out 20 stocks. Along with individual stocks, we also test the performance on *SH Composite Index*, which can be seen as the average performance of all stocks. And we use data between 2015 and 2017 for training, data after 2017 for testing. One training instance

---

[1] http://news.baidu.com/.

[2] http://news.sina.com.cn/.

[3] http://ictclas.nlpir.org/.

Table 1. Statistics of the dataset

Data statistics	Value
Count of events	163867
Avg count of events per day	212
Avg length of event title	13
Avg length of event document	196
Count of companies (stocks)	2964
Avg length of stock meta-information	218
Count of industries	35

consists of stock meta-information, events happen in a specific day and the up or down trend after a fixed period. We set the period to day (short-term), week (medium-term), or month (long-term), and the experiments about this setting is in Sect. 4.4. In medium-term setting, we get a training set of size 511400 and testing set of size 81200.

4.2 Implementation and Hyper-Parameter Setting

The performance of deep learning models dependent on the chosen of hyper-parameters heavily. In our experiments, we tuning our model by use grid search with 5-cross validation to find the best-fitted hyper-parameters.

We initialize word vector values from gaussian distribution, and we set the vector dimension w_d to 200 to achieve the speed and performance balance. For other hyper-parameters, we chose LSTM hidden dimension h_{lstm} from $\{100, 200, 300\}$; CNN window size w_d from $\{2, 3, 4, 5\}$; CNN filter map f_n from $\{64, 128, 256, 512\}$; non-linear hidden dimension h_p from $\{100, 200, 300\}$; learning rate l from $\{0.001, 0.01, 0.1, 1\}$; regularization parameter λ from $\{0.0001, 0.001,$

Table 2. Hyper-parameter setting

Hyper-parameter names	Value
Word embedding w_d	200
LSTM hidden dimension h_{lstm}	300
CNN window size w_d	4
CNN filter maps f_n	128
Non-linear hidden size h_p	200
Learning rate l	0.001
Batch size bz	512
Dropout rate ρ	0.6
Regularization parameter λ	0.1

0.01, 0.1}, and dropout rate ρ from 0.1 to 0.7. We set batch size bz to 512 without tuning because its minor influential to the final performance. The best fitted hyper-parameters are summarized in Table 2.

4.3   Stock Price Movement Prediction Experiments

We compare ATT-RENN model with several baselines, note that all the reported performances are under the medium-term setting, namely, we set the fixed period to one week (In fact, all the models show similar consistent performance with respect to different time periods). We illustrate different models below:

RandomGus is a basic baseline model that ignores all the information totally. It simply outputs 0 (*down*) or 1 (*up*) at a chance of 50%. We also use this model to check the balance of the dataset.

StructureAvg is the method proposed in Ding [7]. Events are extracted from syntax trees by heuristic rules and are represented as $<S, V, O>$. "Avg" means we don't compute relevance for each event but we assume that all events contribute equally for the change of prices.

StructureAtt represents event the same structure as StructureAvg. Additionally, it computes events relevance use attention mechanism and combines information according to events relevance weights.

ATT-ERNNAvg uses LSTM encoder to encode events. Compared with the full ATT-TRNN model, it doesn't compute event relevance, and averaged sum strategy is used.

ATT-ERNNTtl is the full model we proposed in this paper, it only uses news titles to get event representations.

ATT-ERNNTtlCnt is the full model we proposed in this paper, it uses news titles and news contents information to compute event representations.

Performances of all these models are listed in Table 3. We use accuracy as the evaluation metric.

The nearly 50% accuracy of RandomGus shows the non-bias of the dataset. Besides, we can draw three important conclusions:

(1) ATT-ERNN achieves the best result (69.3%) in individual stock price prediction when only using news titles to represent events. It outperforms the model without event relevance computing (ATT-ERNNAvg) by 2.5%.

Table 3. Performance of different models

Models	SH composite index	Individual stocks
RandomGus	49.6	50.7
StructureAvg	63.4	64.9
StructureAtt	63.2	65.6
ATT-ERNNAvg	**66.5**	66.8
ATT-ERNNTtl	65.9	**69.3**
ATT-ERNNTtlCnt	65.7	68.7

For *SH Composite Index*, ATT-ERNNAvg achieves the best result (66.5%). Since *SH Composite Index* can be seen as the **average** of all stocks, which just prove the reasonableness of using attention mechanism in the opposite way.

(2) Model with distributed representations of events gets better performance (+3.7%) compared with model with structure representations. We explain the phenomenon as that using end-to-end method can avoid complex feature engineering and prevents error propagation from existing NLP tools.

(3) Using news titles and news contents gets worse result (−0.6%). We argue that news title is the most informative text to indicate an event occur. And when we take news content into consideration, it might lead in much noise, which might confuse the model and results in worse performance.

From the third conclusion, we get the intuition that adding more data not always brings improvement of performance. To verify this idea, we conduct experiments exploiting the prediction accuracy with respect to the number of events count. We sort all events happen in a day according to their summed TF-IDF values to get the rough order of events by importance, and we take n event from top to down. Experimental results are shown in Fig. 2.

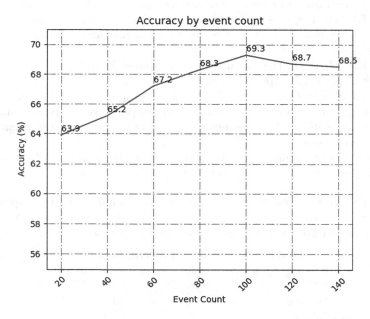

Fig. 2. The performance of our model with respect to event count

As shown in Fig. 2, when data is insufficient (less than 100 events), adding more data gets better performance. However, when it hits a threshold, adding more data won't benefit any more, and the performance even starts to drop.

The results justify our hypotheses that adding more data will also lead much noise, and then the model will have trouble in finding the right patterns in data, which results in poor performance.

4.4 Short-Term, Medium-Term, Long-Term Influence

We illustrate experiments exploiting the influences of short-term (a day), medium-term (a week), and long-term (a month) on performance. We set period as above and rebuild training and testing data, and then we apply ATT-RENN model to these three settings. The experimental results are plotted in Fig. 3.

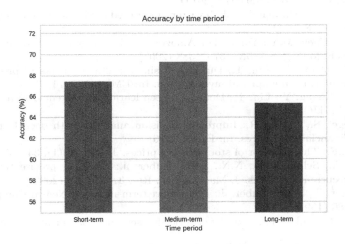

Fig. 3. The performance with respect to time period

From the result, we find that medium-term setting yields out the best performance. We explain the phenomenon as that: short-term setting doesn't provide enough time for the market to react, while when setting to long-term, too many other factors will dominate market trending, which also results in poor performance. Our experiment results show that medium-term, namely one week, is the appropriate period for the market to reflect the effect of events on stock prices.

5 Conclusion and Future Work

In this paper, we propose ATT-ERNN model to exploit the implicit correlations between world events and the movement of stock prices. We conduct extensive experiments on a real-world dataset, and experimental results show the superiority of our model. In the future, we will do fine-grained quantitative analysis and add real market simulation to explore the model's ability further.

Acknowledgement. This work was supported by the Natural Science Foundation of China (No. 61533018) and the National Basic Research Program of China (No. 2014CB340503). And this research work was also supported by Google through focused research awards program.

References

1. Andersen, T.G., Bollerslev, T.: Intraday periodicity and volatility persistence in financial markets. J. Empirical Finan. **4**(2), 115–158 (1997)
2. Batres-Estrada, G.: Deep Learning for Multivariate Financial Time Series (2015)
3. Bahdanau, D., et al.: Neural machine translation by jointly learning to align and translate. CoRR abs/1409.0473 (2014)
4. Bollen, J., Mao, H., Zeng, X.-J.: Twitter mood predicts the stock market. J. Comput. Sci. **2**, 1–8 (2011)
5. Das, S.R., Chen, M.Y.: Yahoo! For Amazon: sentiment extraction from small talk on the web. Manag. Sci. **53**, 1375–1388 (2007)
6. Ding, X., Zhang, Y., Liu, T., Duan, J.: Using structured events to predict stock price movement: an empirical investigation. In: EMNLP (2014)
7. Ding, X., Zhang, Y., Liu, T., Duan, J.: Deep learning for event-driven stock prediction. In: IJCAI (2015)
8. Feuerriegel, S., Fehrer, R.: Improving decision analytics with deep learning: the case of financial disclosures. In: ECIS (2016)
9. Fama, E.F.: The behavior of stock-market prices. J. Bus. **38**(1), 34–105 (1965)
10. Gers, F.A., Schraudolph, N.N., Schmidhuber, J.: Learning precise timing with LSTM recurrent networks. J. Mach. Learn. Res. **3**, 115–143 (2002)
11. Hochreiter, S., Schmidhuber, J.: Long short-term memory. Neural Comput. **9**(8), 1735–1780 (1997)
12. Kim, Y.: Convolutional neural networks for sentence classification. In: EMNLP (2014)
13. Krizhevsky, A., Sutskever, I., Hinton, G.E.: ImageNet classification with deep convolutional neural networks. In: NIPS (2012)
14. Mikolov, T., et al.: Distributed representations of words and phrases and their compositionality. In: NIPS (2013)
15. Malkiel, B.G.: The efficient market hypothesis and its critics. J. Econ. Perspect. **17**(1), 59–82 (2003)
16. Malkiel, B.G.: A Random Walk Down Wall Street: The Time-Tested Strategy for Successful Investing. Princeton University, Princeton (2003)
17. Rnnqvist, S., Sarlin, P.: Bank distress in the news: describing events through deep learning. CoRR abs/1603.05670 (2016)
18. Rush, A.M., et al.: A neural attention model for abstractive sentence summarization. In: MNLP (2015)
19. Srivastava, N., Hinton, G.E., Krizhevsky, A., Sutskever, I., Salakhutdinov, R.: Dropout: a simple way to prevent neural networks from overfitting. J. Mach. Learn. Res. **15**, 1929–1958 (2014)
20. Sutskever, I., Vinyals, O., Le, Q.V.: Sequence to sequence learning with neural networks. In: NIPS (2014)
21. Taylor, S.J., Xu, X.: The incremental volatility information in one million foreign exchange quotations (2003)
22. Tetlock, P.C., Saar-Tsechansky, M., Macskassy, S.: More than words: quantifying language to measure firms fundamentals (1998)

23. Tetlock, P.C.: Giving content to investor sentiment: the role of media in the stock market. J. Finan. **62**(3), 1139–1168 (2007)
24. Takeuchi, L.: Applying deep learning to enhance momentum trading strategies in stocks (2013)
25. Xu, K., Ba, J., Kiros, J.R., Cho, K., Courville, A.C., Salakhutdinov, R., Zemel, R.S., Bengio, Y.: Show, attend and tell: neural image caption generation with visual attention. In: ICML (2015)

A Survey on Relation Extraction

Meiji Cui[1(✉)], Li Li[1], Zhihong Wang[2], and Mingyu You[1]

[1] Tongji University, Shanghai 201804, China
cui_mj@163.com
[2] Shanghai Hi Knowledge Information Technology Ltd., Shanghai 200433, China

Abstract. Relation extraction, as an important part of information extraction, can be used for many applications such as question-answering and knowledge base population. To thoroughly comprehend relation extraction, the paper reviews it mainly concentrating on its mainstream methods. Besides, open information extraction (OIE), as a different relation extraction paradigm, is introduced as well. Also, we exploit the challenges and directions for relation extraction. We hope the paper will give the overview of relation extraction and help guide the path ahead.

Keywords: Relation extraction · Distant supervision · Neural network · OpenIE
End-to-end extraction

1 Introduction

Relation extraction (RE) is the task of detecting and classifying predefined relationships between entities identified in text. The key objective of the RE is to extract tuples of the form relation <entity1, entity2>. Many efforts have been invested in RE, and there are many varied existing approaches to this problem. At present, the mainstream approaches of RE can be classified as follows: rule-based methods and statistic-based methods, while statistic-based methods include unsupervised approaches, semi-supervised approaches, supervised approaches, distant supervision, neural network. We will discuss these approaches in detail in subsequent sections.

The rest of the paper is organized as follows: Sect. 2 introduces some datasets which are used in the task of RE. The existing mainstream approaches for RE are summarized in Sect. 3. Open information extraction is demonstrated in Sect. 4. Challenges and possible directions are depicted in Sect. 5 and some conclusions are drawn in Sect. 6.

2 Datasets for Relation Extraction

The previous works on RE mainly employed supervised training datasets. These datasets are intensively human annotated. ACE 2004 dataset, ACE 2005 dataset and SemEval-2010 Task 8 are the main supervised datasets for RE.

Since the human-annotated datasets for relation extraction are time-consuming and generally small, Mintz et al. proposed a distant supervision approach for automatically

© Springer Nature Singapore Pte Ltd. 2017
J. Li et al. (Eds.): CCKS 2017, CCIS 784, pp. 50–58, 2017.
https://doi.org/10.1007/978-981-10-7359-5_6

generating large amounts of training data [1]. Normally, Wikipedia, NY Times corpus [2] and KBP shared tasks [3] are used as evaluation datasets for relation extraction.

3 Mainstream Methods

Since RE plays a vital role in robust knowledge extraction from unstructured texts and serves as an intermediate step in a variety of natural language processing applications, many efforts have been invested in it. Some mainstream RE approaches will be depicted as follows.

3.1 Rule-Based Approaches

The most previous works on RE are based on rules. Rule-based approaches need to predefine rules that describe the structure of the entity mentions. These methods require that the rules builder have a deep understanding of the background and characteristics of the field. Hence, the obvious drawbacks are the huge demand of human participation and poor portability.

3.2 Statistic-Based Approaches

Nowadays, statistic-based approaches can be simply classified into five categories: unsupervised, semi-supervised, supervised, distant supervision and neural network. The details of such approaches will be demonstrated.

Unsupervised Approaches. Unsupervised methods extract strings of words between entities in large amounts of text, and clusters and simplifies these word strings to produce relation-strings [4]. Without demand of annotated data, unsupervised methods can use very large amounts of data as well as extract very large amount of relations. However, since there is no standard form of relations, the output resulting may not be easy to map to relations which is necessary for a particular knowledge base.

Semi-supervised Approaches. The main idea of semi-supervised methods is bootstrapping, thereby the methods are also known as bootstrapping methods. This was first introduced in DIPRE [5], and then extended in Snowball, KnowltAll and TextRunner. Bootstrapping methods typically suffer from semantic drift and poor precision.

Supervised Approaches. Supervised approaches are the most commonly used methods for relation extraction and yield relatively high performance. In the supervised paradigm, relation extraction is treated as a classification task. Supervised method can be simply divided into two types: feature-based methods and kernel-based methods.

In feature-based methods, classification clues (e.g., sequences, parse trees) need to be exploited to convert into feature vectors [6]. The key of feature-based approaches is how to select a suitable feature set when converting structured representations into feature vectors. However, choice of features is usually guided by intuition and experiments.

Kernel-based methods provide a natural alternative to exploit rich representations of input classification clues, such as syntactic parse trees. Kernel defines similarity between objects (e.g. strings, word sequences and parse trees) implicitly in a higher dimensional space. Reference [7] proposed the shortest-path dependency kernels. Lately, a new feature-enriched tree kernel for relation extraction was proposed, by annotating each tree node with a set of discriminate features to refine the syntactic tree representation [8].

However, supervised methods rely on the availability of extensive training data. Additionally, because the relations are labeled on a particular corpus, the resulting classifiers are of poor portability toward other domains. What's more, some NLP tools used in supervised methods are prone to produce some errors, harmful to relation extraction.

Distant Supervision. Distant supervision automatically generates training examples and learns features through aligning free text with KBs such as Freebase, a large semantic database. Thus the method does not need any human intervention, and can extract vast numbers of features from a large amount of data. Table 1 represents the main development process of distant supervision.

Table 1. Distant supervision based methods for RE

Model/System	Main points	Addressed problems
Logistic classifier [1]	DS for RE SISL	Manual annotation
At-Least-One Model [2]	Relaxing SISL to MISL	Predicting relations
MultiR [10]	MIML two approximations	Overlapping relations
MIML-RE [3]	Graphical model	Overlapping relations

The method was first introduced in the context of biological KBs [9]. With the availability of large scale of KBs such as Freebase, Reference [1] extended distant supervision method to any texts. They made the assumption that if two entities participate in a relation, any sentence that contains those two entities might express that relation (SISL). However, the assumption is too strong, treating the relation extraction as a single-instance single-label task. Hence, Reference [2] relaxed the assumption which is called expressed-at-least-once assumption holding with more certainty (e.g. multi instances, MISL). Reference [10] further relaxed the consumption by constructing a graphical model, treating the relation extraction as a multi-instance multi-label problem (e.g. MIML). Reference [3] used a multi-label classifier to denote the latent relation types of all the mentions involving that pair and a set of binary classifiers to decide if the relation holds for the given entity pair.

The above methods ignored the key point that knowledge bases used as distant supervision were highly incomplete. Therefore, they proposed a semi-supervised MIML algorithm modeling the bag-level label noise where entity-pair level labels were either positive or unlabeled. Reference [11] also extended the model to resolve the incompleteness of knowledge bases. However, these methods only exploit one specific kind of indirect supervision knowledge, but ignore many other kinds of useful supervision knowledge. Thus, some distant supervision methods by using additional knowledge to eliminate wrongly labeled instances were proposed [12].

Neural Network. All of the above methods' performance strongly depends on the quality of the extracted features, derived from the existing natural language processing (NLP) tools. As we all know, the errors are inevitably produced during the processing. Hence, how to extract features by reducing the use of existing NLP tools becomes the important research point. The resurgence of neural network (NN) provides the new insight to such problem. Neural network was first applied to relation classification by Reference [13]. Since then, many neural networks, such as RNN, CNN (convolution neural network), LSTM (long short term memory), are exploited to relation extraction.

Zeng et al. employed a CNN-based framework for relation extraction, which was the first attempt [14]. They employed a CNN to extract both lexical and sentence level features, and then concatenated to form the final feature vector to predict the relation between entity pair. Reference [15] used multiple window sizes and pre-trained word embeddings to improve the performance of CNN-based model. To deal with the impact of artificial classes, Reference [16] also adopted a CNN-based model, using a new pairwise ranking loss function to perform relation classification by ranking (CR-CNN). Xu et al. acquired relation representations from shortest dependency path through a CNN [17]. Reference [18] further combined distant supervision and piecewise CNN (PCNN). Additionally, Jiang et al. considered different sentences contained the entity pair by performing cross-sentence max-pooling to select features from different sentences of interest after extracting each sentence of interest based on a CNN [19]. Reference [20] took attention mechanism to relation extraction by considering more sentences of interest in order to make full use of supervision information. Attention mechanism aimed

Table 2. CNN-based methods for relation extraction

Classifier	Additional features	Main points
MVRNN [13]	Syntactic parsing tree, POS, NER, WordNet	Matrix-vector representation
CNN [14]	PF, WordNet, words around nominals	PF
CR-CNN [16]	PF	Ranking approach
depLCNN + NS [17]	Dependency parsing, WordNet, words around nominals	DSP + Negative sampling strategy
PCNN [18]	PF	Piecewise max pooling
MIMLCNN [19]	PF	Cross-sentence max pooling, multi-label relation modeling
ACNN/APCNN [20]	PF	Attention mechanism
Att-Pooling-CNN [21]	PF	Attention both on entity level and pooling level
Path + Max [22]	PF	Inference chains built on relation path

to automatically increase the weights of invalid instances while reducing the weights of those noisy instances. A CNN with two levels of attention was adopted in Reference [21]. To address the problem that many sentences containing only one target entity while rich in information for RE, Reference [22] originally incorporated relation path into RE. Table 2 shows the CNN-based methods for RE. Since word embeddings are all used in neural network, they don't be considered as additional features.

However, CNN-based method can't capture temporal features, crucial for the performance of RE while the distance between target entities is long. Another neural network called recurrent neural network (RNN) was used in RE for that purpose. Reference [23] proved that RNN particularly performed better in learning relations within long context, attributing to the bi-directional network. The long short term memory (LSTM) architecture extended by RNN was adopted by Reference [24] to model the complete sentence. Xu et al. combined shortest dependency path (SDP) with LSTM units, while LSTM networks were multichannel corresponding to different types of information along the SDP, allowing heterogeneous sources integrated [25]. Similar to the CNN model, attention mechanism is also used in LSTM, which was proposed by Reference [26]. Table 3 represents the methods based on RNN for relation extraction.

Table 3. RNN-based methods for relation extraction

Classifier	Additional features	Main points
RNN [23]	Position indicator	Simple RNN model
BLSTM [24]	PF, POS, NER, WNSYN, DEP, RELATIVE-DEP	BLSTM model
SDP + LSTM [25]	Dependency parsing, POS, GR, WordNet	Combination of SDP and LSTM
Att + BLSTM [26]	PI	Attention mechanism

All above methods treated entity and relation extraction as a pipeline of two separated tasks, i.e., named entity recognition (NER) and relation classification (RC). However, the two sub-tasks are much relevant such that joint model can avoid cascading of errors. Therefore, end-to-end relation extraction method is proposed, which infers to modeling entities and relations in a single model. Reference [27] firstly used a single model to jointly predict entities and relations by an incremental beam-search algorithm in conjunction with structured perceptron. Reference [28] used a history-based structured learning approach to jointly extract entities and relations. Reference [29] utilized bi-directional LSTM-RNNs on both word sequence level and dependency tree substructure information level, while stacking tree-structured LSTM-RNNs on sequential LSTM-RNNs, allowing the two LSTM-RNNs sharing parameters in a model. Reference [30] further proposed a hybrid neural network, while the first layer is a bi-directional LSTM unit, shared by both NER module and RC module. Then the second layer includes a LSTM unit to recognize entities, which can capture tag dependencies and a CNN layer to extract relations.

4 Open Information Extraction (OIE)

Compared with the above methods, open information extraction (OIE), originally introduced by Reference [31] called TextRunner, aims to obtain tuples with highly scalable extraction in portable across domain by identifying a variety of relation phrases and their arguments in arbitrary sentences. Reference [32] presented a novel OIE system called WOE which means Wikipedia-based open extractor. The training data they used were produced by heuristically matching Wikipedia infobox attribute values with corresponding sentences. Additionally, they also proved that dependency parse features performed better than shallow linguistic features. Furthermore, Reference [33] proposed the second generation of OIE to deal with the problem of incoherent extractions and uninformative extractions, by introducing REVERB which implements a general model of verb-based relation phrases expressed as two simple constraints, e.g. syntactic constraint and lexical constraint. Also, they proposed R2A2 adding an argument learning component, ARGLEARNER, which performed significantly better than the previous work. Reference [34] introduced a model called OLLIE, e.g. open language learning for information extraction. Reference [35] proposed an OIE system called ClausIE, aiming at solving the problems of relations expressed via appositions, possessives and participial modiers.

5 Challenges and Directions

Relation extraction, as a key component of information extraction, is still a challenging research field due to the complexity of human language.

For distant supervision, KBs are still far away from the total completed ones. On the one hand, we can complete the KBs as possible, which can be used as powerful distant supervision for relation extraction. On the other hand, some effective inference chains can be considered to extract more relations while no completed entity mentions in KBs.

As we can see, neural network is widely used in relation extraction and has achieved comparable results. We can make progress is to make full use of neural network with little manual features or features extracted by NLP toolkits. Hence, we should fully extract features automatically from all possible instances of interest with neural network. The above mentioned attention mechanism is the promising method for relation extraction together with neural network. We can exploit more levels attention to deal with the multi-instances multi-relations problem.

Regarding that entities extraction and relations extraction influence each other. Joint extraction can be seen as the promising method. We can use more effective or efficient methods, such as method used in Reference [36] for NER, to improve the performance.

6 Conclusion

In this paper, we review the relation extraction in a system. We include the mainstream relation extraction methods, rule-based methods and statistic-based methods respectively. OpenIE, as a different relation extraction paradigm, is introduced as well.

Particularly, we exploit the challenges and directions for relation extraction based on distant supervision and neural network. We can see, relation extraction, as an important part for question-answering, knowledge base population and so on, is a promising and meaningful research field as it is still faced with many challenges.

Acknowledgment. This research is supported by the National Natural Science Foundation of China under Grant No. 51475334, the Science and Technology Project of Shanghai under Grant No. 16dz1206102 and the Fundamental Research Funds for the Central Universities under Grant No.22120170077.

References

1. Mintz, M., Bills, S., Snow, R., et al.: Distant supervision for relation extraction without labeled data. In: Proceedings of the Joint Conference of the 47th Annual Meeting of the ACL and the 4th International Joint Conference on Natural Language Processing of the AFNLP, vol. 2, pp. 1003–1011 (2009)
2. Riedel, S., Yao, L., McCallum, A.: Modeling relations and their mentions without labeled text. In: Balcázar, J.L., Bonchi, F., Gionis, A., Sebag, M. (eds.) ECML PKDD 2010. LNCS (LNAI), vol. 6323, pp. 148–163. Springer, Heidelberg (2010). https://doi.org/10.1007/978-3-642-15939-8_10
3. Surdeanu, M., Tibshirani, J., Nallapati, R., et al.: Multi-instance multi-label learning for relation extraction. In: Proceedings of the 2012 Joint Conference on Empirical Methods in Natural Language Processing and Computational Natural Language Learning, pp. 455–465 (2012)
4. Yan, Y., Okazaki, N., Matsuo, Y., et al.: Unsupervised relation extraction by mining Wikipedia texts using information from the web. In: Proceedings of the Joint Conference of the 47th Annual Meeting of the ACL and the 4th International Joint Conference on Natural Language Processing of the AFNLP, vol. 2, pp. 1021–1029 (2009)
5. Brin, S.: Extracting patterns and relations from the world wide web. In: Atzeni, P., Mendelzon, A., Mecca, G. (eds.) WebDB 1998. LNCS, vol. 1590, pp. 172–183. Springer, Heidelberg (1999). https://doi.org/10.1007/10704656_11
6. Kambhatla N.: Combining lexical, syntactic, and semantic features with maximum entropy models for extracting relations. In: Proceedings of the ACL 2004 on Interactive Poster and Demonstration Sessions (2004)
7. Bunescu R.C., Mooney R.J.: A shortest path dependency kernel for relation extraction. In: Proceedings of the Conference on Human Language Technology and Empirical Methods in Natural Language Processing, pp. 724–731 (2005)
8. Sun, L., Han, X.: A feature-enriched tree kernel for relation extraction (2014)
9. Craven, M., Kumlien, J.: Constructing biological knowledge bases by extracting information from text sources. In: Proceedings of ISMB, pp. 77–86 (1999)
10. Hoffmann, R., Zhang, C., Ling, X., et al.: Knowledge-based weak supervision for information extraction of overlapping relations. In: Proceedings of the 49th Annual Meeting of the Association for Computational Linguistics, vol. 1, pp. 541–550 (2011)
11. Ritter, A., Zettlemoyer, L., Etzioni, O.: Modeling missing data in distant supervision for information extraction. Trans. Assoc. Comput. Linguist. **1**, 367–378 (2013)
12. Han, X., Sun, L.: Global distant supervision for relation extraction. In: Proceedings of Thirtieth AAAI Conference on Artificial Intelligence (2016)

13. Socher, R., Huval, B., Manning, C.D., et al.: Semantic compositionality through recursive matrix-vector spaces. In: Proceedings of the 2012 Joint Conference on Empirical Methods in Natural Language Processing and Computational Natural Language Learning, pp. 1201–1211 (2012)

14. Zeng, D., Liu, K., Lai, S., et al.: Relation classification via convolutional deep neural network. In: Proceedings of COLING, pp. 2335–2344 (2014)

15. Nguyen, T.H., Grishman, R.: Relation extraction: perspective from convolutional neural networks. In: Proceedings of VS@ HLT-NAACL, pp. 39–48 (2015)

16. Santos, C.N., Xiang, B., Zhou, B.: Classifying relations by ranking with convolutional neural networks. arXiv preprint arXiv:1504.06580 (2015)

17. Xu, K., Feng, Y., Huang, S., et al.: Semantic relation classification via convolutional neural networks with simple negative sampling. arXiv preprint arXiv:1506.07650 (2015)

18. Zeng, D., Liu, K., Chen, Y., et al.: Proceedings of the 2015 Conference on Empirical Methods in Natural Language Processing, pp. 1753–1762 (2015)

19. Jiang, X., Wang, Q., Li, P., et al.: Relation extraction with multi-instance multi-label convolutional neural networks. In: COLING, pp. 1471–1480 (2016)

20. Lin, Y., Shen, S., Liu, Z., et al.: Neural relation extraction with selective attention over instances. In: Proceedings of ACL, vol. 1 (2016)

21. Wang, L., Cao, Z., de Melo, G., et al.: Relation classification via multi-level attention CNNs. In: Proceedings of the 54th Annual Meeting of the Association for Computational Linguistics, vol. 1, pp. 1298–1307 (2016)

22. Zeng, W., Lin, Y., Liu, Z., et al.: Incorporating relation paths in neural relation extraction. arXiv preprint arXiv:1609.07479 (2016)

23. Zhang, D., Wang, D.: Relation classification via recurrent neural network. arXiv preprint arXiv:1508.01006 (2015)

24. Zhang, S., Zheng, D., Hu, X., et al.: Bidirectional long short-term memory networks for relation classification. In: 29th Pacific Asia Conference on Language, Information and Computation, pp. 73–78 (2015)

25. Xu, Y., Mou, L., Li, G., et al.: Classifying relations via long short term memory networks along shortest dependency paths. In: Proceedings of the 2015 Conference on Empirical Methods in Natural Language Processing, pp. 1785–1794 (2015)

26. Zhou, P., Shi, W., Tian, J., et al.: Attention-based bidirectional long short-term memory networks for relation classification. In: Proceeding of the 54th Annual Meeting of the Association for Computational Linguistics, pp. 207–212 (2016)

27. Li, Q., Ji, H.: Incremental joint extraction of entity mentions and relations. In: Proceedings of ACL, vol. 1, pp. 402–412 (2014)

28. Miwa, M., Sasaki, Y.: Modeling joint entity and relation extraction with table representation. In: Proceedings of EMNLP, pp. 1858–1869 (2014)

29. Miwa, M., Bansal, M.: End-to-end relation extraction using LSTMs on sequences and tree structures. arXiv preprint arXiv:1601.00770 (2016)

30. Zheng, S., Hao, Y., Lu, D., et al.: Joint entity and relation extraction based on a hybrid neural network. Neurocomputing 257(27), 59–66 (2017)

31. Yates, A., Cafarella, M., Banko, M., et al.: TextRunner: open information extraction on the web. In: Proceedings of Human Language Technologies: The Annual Conference of the North American Chapter of the Association for Computational Linguistics: Demonstrations, pp. 25–26 (2007)

32. Wu, F., Weld, D.S.: Open information extraction using Wikipedia. In: Proceedings of the 48th Annual Meeting of the Association for Computational Linguistics, pp. 118–127 (2010)

33. Etzioni, O., Fader, A., Christensen, J., et al.: Open information extraction: the second generation. IJCAI **11**, 3–10 (2011)

34. Schmitz, M., Bart, R., Soderland, S., et al.: Open language learning for information extraction. In: Proceedings of the 2012 Joint Conference on Empirical Methods in Natural Language Processing and Computational Natural Language Learning, pp. 523–534 (2012)

35. Del Corro, L., Gemulla, R.: ClausIE: clause-based open information extraction. In: Proceedings of the 22nd International Conference on World Wide Web, pp. 355–366 (2013)

36. Ma, X., Hovy, E.: End-to-end sequence labeling via bi-directional LSTM-CNNs-CRF. arXiv preprint arXiv:1603.01354 (2016)

A Sentiment and Topic Model with Timeslice, User and Hashtag for Posts on Social Media

Kang Xu$^{(\boxtimes)}$, Junheng Huang, and Tianxing Wu

Department of Computer Science, Southeast Unversity, Nanjing, China
{kxu,jhhuang,wutianxing}@seu.edu.cn

Abstract. Nowadays plenty of user-generated posts, e.g., tweets, are published on social media and the posts imply the public's opinions towards various topics. Joint sentiment/topic models are widely applied in detecting sentiment-aware topics on the lengthy documents. However, the characteristics of posts, i.e., short texts, on social media pose new challenges: (1) context sparsity problem of posts makes traditional sentiment-topic models infeasible; (2) conventional sentiment-topic models are designed for flat documents without structure information, while publishing users, publishing timeslices and hashtags of posts provide rich structure information for modeling these posts. In this paper, we firstly devise a method to mine potential hashtags, based on explicit hashtags, to further enrich structure information for posts, then we propose a novel Sentiment Topic Model for Posts (STMP) which aggregates posts with the structure information, i.e., timeslices, users and hashtags, to alleviate the context sparsity problem. Experiments on Twitter7 show STMP outperforms previous models in sentiment-aware topic extraction.

Keywords: Topic model · Sentiment analysis · Topic extraction
Short text

1 Introduction

With the rapid growth of Web 2.0, a mass of user-generated posts, e.g., tweets, which capture people's interests, thoughts, sentiments and actions. The posts have been accumulating on the social media with each passing day. Sentiment analysis attempts to find user preference, likes and dislikes from the posts on social media, such as reviews and microblogs [5] and topic modeling attempts to discover the topics or aspects from reviews and microblogs [1]. Topic modeling and sentiment analysis on the posts are two significant tasks which can benefit many people. Topic modeling and sentiment analysis on the social media are complementary where sentiments on the social media often change over different topics and topics on the social media are always related to public sentiments. So jointly modeling topics and sentiments on the social media is a significative task and it can reflect people's sentiments on different topics, e.g., a topic about "Apple Inc." ('ipad', 'iphone', 'itouch', 'imac', 'beautiful' and 'popular') with

© Springer Nature Singapore Pte Ltd. 2017
J. Li et al. (Eds.): CCKS 2017, CCIS 784, pp. 59–65, 2017.
https://doi.org/10.1007/978-981-10-7359-5_7

the overall sentiment polarity "positive". Conventional sentiment-aware topic models, like Joint Sentiment/Topic Model (JST) [4] and Aspect/Sentiment Unification Model (ASUM) [3], are utilized for uncovering the hidden topics and sentiments on lengthy documents without structure information. However, the characteristics of posts, i.e., short texts, on social media pose both a challenge and an opportunity: (1) context sparsity problem of posts makes traditional sentiment-topic models infeasible; (2) conventional sentiment-topic models are designed for documents without structure information, while users and timeslices of posts and hashtags contained in posts provide structure information for posts.

One simple and effective way to alleviate the sparsity problem is to aggregate short posts into lengthy pseudo-documents [2,9,10]. Inspired by the observations mentioned above, Xu et al. proposes a Time-User Sentiment/Topic Latent Dirichlet Allocation (TUS-LDA) [8]. TUS-LDA is based on the assumption that the posts on the social media are a mixture of two kinds of topics: temporal topics which are related to current events and stable topics which are related to personal interests. TUS-LDA takes advantage of structure information of posts, i.e., users (who publish posts) and timeslices (when posts are published), to aggregate posts for alleviating context sparsity problems of posts.

Moreover, we observe that hashtags, prefixing one or more characters with a hash symbol as "#hashtag", provide another kind of structure information for posts. Semantic relations between posts are built with hashtags, where posts under the same hashtags always talk about the similar topics. Since hashtags are strong topic indicators labeled by users on social media. Hence, hashtags are more effective, than timeslices and users, to be utilized for aggregating short texts on social media into pseudo-documents. We also find that there exist two kinds of hashtags for posts: explicit hashtags which are explicitly contained in the posts and potential hashtags which are not explicitly contained in posts but have semantic relevance with explicit hashtags. [7] proposes a method to mine potential hashtags, based on explicit hashtags, which exploit the co-occurrences of hashtags. For example, {D1: Healthy lunch if egg and broccoli **#cooking#food**; D2: Testing **#recipes** in my kitchen all day. I hated **#cooking**; D3: Wonderful day. nice movie thanks **#tweet**; D4: Powers of sping vegetable with chicken enrich breakfast **#recipes#tweet**}, "D1" has "#cooking" and "#food", "D2" has explicit hashtags "#cooking" and "#recipes"; for the occurrences of ("#cooking", "#recipes"), ("#cooking", "#food"), "D1" has a potential hashtag "#recipes" and "D2" has a potential hashtag "#food". However, the directly co-occurred hashtags of explicit hashtags contain many noisy hashtags that are not relevant to explicit hashtags. For example, "D3" introduces a wrong hashtag "#recipes". To optimize the introduction of potential hashtags, embedding representations of all the words (including hashtags) in the posts are learnt using vector arithmetic, such as Word2Vec [6]. Fine-grained semantic and syntactic regularities of hashtags are captured, where semantically close hashtags are adjacent on embedding space. Hence, cosine similarities are utilized to compute the semantic distances of hashtags to sift out potential hashtags for each hashtag.

In this paper, we propose a novel Sentiment Topic Model for Posts (STMP) which aggregates posts with structure information, i.e., timeslices, users or hashtags, to model topics and sentiments for posts on social media. Firstly, we design a simple and effective method to mine potential hashtags for all the explicit hashtags and a new model, STMP, which aggregates posts in the same timeslice, user or hashtag as a pseudo-document to alleviate the context sparsity problem. Then, we design approaches of parameter inference and incorporating prior sentiment knowledge for STMP. Finally, we implement experiments on Tweet7 to evaluate the effectiveness of topic extraction in STMP.

2 Sentiment Topic Model for Posts

Sentiment Topic Model for Posts (STMP) is a probabilistic generative model that describe a process of generating posts on social media with structure information from users, timeslices, explicit and potential hashtags. When a user u_d publishes a post p_d within a timeslice t_d and explicit hashtags H_d, if the post contains at least one hashtag ($|H_d| > 0$), the user firstly chooses a key hashtag h_d^1 from explicit hashtags H_d, then utilize the variable x_d, which is drawn from τ, to decide how to assign hashtags in post d; if $x_d = 0$, the representative h_d is set as the key hashtag h_d^1; otherwise, the user chooses h_d from $R(h_d^1)$, where $R(h_d^1)$ are the potential hashtags of h_d^1; if the post does not contain any hashtsgs ($|H_d| = 0$), the user first utilizes the variable y_d, which is drawn from the global user-timeslice switch distribution ε, to decide whether the post talks about a stable topic or a temporal topic. Then the user chooses a sentiment label l_d for the post from the document-sentiment π_d. If the user chooses a hashtag topic h_d and a sentiment label l_d, the user then selects a topic z_d from o_{h_d,l_d}; if the user chooses a stable topic u_d and a sentiment label l_d, the user then selects a topic z_d from δ_{u_d,l_d}; otherwise, the user selects a topic z_d from θ_{t_d,l_d}. For each word $w_{d,i}$ in the post p_d, the user first chooses a sentiment label $l_{d,i}$; with the chosen topic z_d and sentiment label $l_{d,i}$, the word is drawn from the sentiment-topic word distribution $\varphi_{l_{d,i},z_d}$. Formally, the generative story for each post is as follows:

1. Draw $\varepsilon \sim Beta(\gamma)$
2. For each timeslice $t = 1, ..., T$
   i. For each sentiment label $s = 0, 1, 2$
      a. Draw $\theta_{t,s} \sim \mathrm{Dir}(\alpha)$
3. For each user $u = 1, ..., U$
   i. For each sentiment label $s = 0, 1, 2$
      a. Draw $\delta_{u,s} \sim \mathrm{Dir}(\alpha)$
4. For each hashtag $h = 1, ..., H$
   i. For each sentiment label $s = 0, 1, 2$
      a. Draw $o_{h,s} \sim \mathrm{Dir}(\alpha)$
5. For each sentiment label $s = 0, 1, 2$
   i. For each topic $k = 1, ..., K$
      a. Draw $\varphi_{s,k} \sim \mathrm{Dir}(\beta)$

6. For each post p_d, $d = 1, ..., D$

    i. Draw $\pi_d \sim \text{Dir}(\lambda)$

    ii. Draw $l_d \sim \text{Multi}(\pi_d)$

    iii. if $|H_d| > 0$, Draw $h_d^1 \sim \text{Uniform}(H_d)$; Draw $x_d \sim Bernoulli(\tau)$; if $x_d = 0$, $h_d = h_d^1$, else, Draw $h_d \sim \text{Multi}(R(h_d^1))$; Draw $z_d \sim \text{Multi}(o_{h_d,l_d})$

    iv. if $|H_d| = 0$, Draw $y_d \sim Bernoulli(\varepsilon)$; if $y_d = 0$, Draw $z_d \sim \text{Multi}(\theta_{u_d,l_d})$, else Draw $z_d \sim \text{Multi}(\delta_{t_d,l_d})$

    v. For each word w $i = 1, ... N_d$

        a. Draw $l_{d,i} \sim \text{Multi}(\pi_d)$

        b. Draw $w_{d,i} \sim \text{Multi}(\varphi_{z_d,l_{d,i}})$

3 Experiment

3.1 Dataset Description and Parameter Settings

For experiments, we performed sentiment-aware topic discovery on tweets, which are characterized by their limited 140 characters text. We selected tweets, which are related to electronic products such as camera and mobile phones, from Tweet7[1]. These tweets contain the description and reviews of various electronic products and correspond to multiple sentiment-aware topics. Besides, each tweet contains the content, the release timeslice, the user information.

To optimize the number of topics K, we empirically ran the models with four values of K: 10, 20, 50 and 100 in Sentiment140 and ran the models with three values of K: 10, 20, 50 in Twitter7 (In Twitter7, these tweets only contain a small number of electronic product-related topics). In our model, we simply selected symmetric Dirichlet prior vectors as is empirically done in JST and ASUM. For JST, ASUM and TS, $\alpha = \frac{50}{K}, \beta = 0.01$ and $\gamma = 0.01$. For STMP and TUS-LDA, we set $\alpha = 0.5, \gamma = 0.01, \tau = 0.01, \lambda = 0.01$ and $\beta = 0.01$. In all the methods, Gibbs sampling was run for 1,000 iterations with 200 burn-in periods.

3.2 Topic Coherence

As our objective is to discover more coherent sentiment-aware topics, so we evaluated the topics manually which is based on human judgement. Figure 1(a) shows that STMP can discover more *coherent* topics [8] than JST, ASUM, TS, TUS-LDA. Thereinto, TUS-LDA and STMP can also discover the nearly equal number of positive and negative topics. Figure 1(b) gives the average *Precision*@20 (proportion of *correct* words [8] in top 20 words) of all coherent topics. STMP performed better than other four models and performed best in $K = 10$. From the above, we can observe that aggregating posts in the same timeslice, user or hashtag as a single document can indeed improve the performance in sentiment classification and sentiment-aware topic extraction in user-generated posts, i.e., STMP consistently outperformed the baseline models.

---

[1] https://snap.stanford.edu/data/twitter7.html.

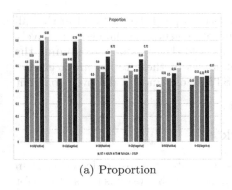

(a) Proportion

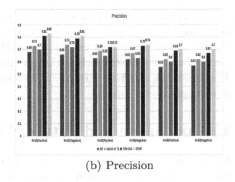

(b) Precision

Fig. 1. (a) Proportion of *coherent* topics generated by each model in $K = 10, 20, 50$ (b) Average Precision @20 (p @20) of words in *coherent* topics generated by each model in $K = 10, 20, 50$

3.3 Topic Visualization

To investigate the quality of topics discovered by STMP, we randomly choose some topics for visualization. We randomly selected six topics, i.e., three positive topics and three negative topics. For each topic, we choose the top 10 words which can most represent the topic. Table 1 presents the top words of the selected topics. The three topics with a positive sentiment label respectively talk about "Camera", "apple music product" and "game" and these topics are listed in the left columns of Table 1; the three negative topics are related to "printer", "window product" and "phone" are listed in right columns of Table 1. As we can see clearly from Table 1, the six topics are quite explicit and coherent, where each of them tried to capture the topic of a kind of electronic product. In terms of

Table 1. Example of topics extracted by STMP

Positive sentiment label			Negative sentiment label		
Topic 1	Topic 2	Topic 3	Topic 1	Topic 2	Topic 3
camera	ipod	xbox	printer	window	phone
digit	song	game	ink	vista	**problem**
canon	phone	live	print	us	information
nikon	listen	sale	cartridge	microsoft	security
new	music	console	**low**	install	**strange**
len	love	microsoft	laser	download	**risk**
photograph	tone	play	color	software	finance
sharp	play	playstate	laserjet	file	mobile
panason	shuffle	ps3	paper	**slow**	digit
slr	**good**	**new**	scanner	server	on-line

topic sentiment, by checking each of the topics in Topic Table 1, it is clear that all the 6 topics can indeed bear positive and negative sentiment labels respectively where all the sentiment words are written in bold. By manually examining the tweet data, we observe that the sentiment labels of these topics are accurate. The analysis of these topics shows that STMP can indeed discover coherent sentiment-aware topics.

4 Conclusion and Future Work

In this paper, we studied the problem of sentiment-aware topic detection from the user-generated posts on the social media. The existing work is not suitable for the short and informal posts, we proposed a new sentiment/topic model for posts on social media that considers the time, user and hashtag information of posts to jointly model topics and sentiments. Based on the different characteristics of sentiments and topics, we limited that words in the same post belong to the same topic, but they can belong to different sentiments. We compared our model with JST, ASUM, TS and TUS-LDA on Tweet7. We asked two judges to evaluate our models and baseline methods and the result also showed that our model STMP performed best in sentiment-aware topic extraction. Moreover, we also chose six examples to visualize some sentiment-aware topics. In the future work, we will incorporate word embedding into our model to improve the performance of modeling posts on social media.

Acknowledgements. This work is supported in part by the National Natural Science Foundation of China (NSFC) under Grant No. 61672153, the 863 Program under Grant No. 2015AA015406 and the Fundamental Research Funds for the Central Universities and the Research Innovation Program for College Graduates of Jiangsu Province under Grant No. KYLX16_0295.

References

1. Chen, Z., Mukherjee, A., Liu, B., Hsu, M., Castellanos, M., Ghosh, R.: Leveraging multi-domain prior knowledge in topic models. In: Proceedings of IJCAI, pp. 2071–2077. AAAI (2013)
2. Diao, Q., Jiang, J., Zhu, F., Lim, E.-P.: Finding bursty topics from microblogs. In: Proceedings of ACL, pp. 536–544. ACL (2012)
3. Jo, Y., Oh, A.H.: Aspect and sentiment unification model for online review analysis. In: Proceedings of WSDM, pp. 815–824. ACM (2011)
4. Lin, C., He, Y.: Joint sentiment/topic model for sentiment analysis. In: Proceedings of CIKM, pp. 375–384. ACM (2009)
5. Mukherjee, S., Basu, G., Joshi, S.: Joint author sentiment topic model. In: SDM, pp. 370–378. SIAM (2014)
6. Rong, X.: Word2Vec parameter learning explained. arXiv preprint arXiv:1411.2738 (2014)
7. Wang, Y., Liu, J., Huang, Y., Feng, X.: Using hashtag graph-based topic model to connect semantically-related words without co-occurrence in microblogs. IEEE Trans. Knowl. Data Eng. **28**(7), 1919–1933 (2016)

8. Xu, K., Qi, G., Huang, J., Wu, T.: A joint model for sentiment-aware topic detection on social media. In: Proceedings of ECAI, pp. 338–346. IOS Press (2016)
9. Zhang, Q., Gong, Y., Sun, X., Huang, X.: Time-aware personalized hashtag recommendation on social media. In: Proceedings of COLING, pp. 203–212. ACL (2014)
10. Zhao, W.X., Jiang, J., Weng, J., He, J., Lim, E.-P., Yan, H., Li, X.: Comparing twitter and traditional media using topic models. In: Clough, P., Foley, C., Gurrin, C., Jones, G.J.F., Kraaij, W., Lee, H., Mudoch, V. (eds.) ECIR 2011. LNCS, vol. 6611, pp. 338–349. Springer, Heidelberg (2011). https://doi.org/10.1007/978-3-642-20161-5_34

Collective Entity Linking Based on DBpedia

Guidong Zheng, Ming Liu[✉], and Bingquan Liu

School of Computer Science and Technology, Harbin Institute of Technology, Harbin, China
{gdzheng,liubq}@insun.hit.edu.cn, liuming1981@hit.edu.cn

Abstract. With the rapid development of Internet, lots of web data are published by internet users. This situation causes tremendous entities appear on the web. However, because of variety and ambiguity of natural language, one entity usually has multiple expressions. To know the actual meaning of one document, it is important to solve the problem of entity ambiguity. Entity linking is a good solution for entity disambiguation. It links one entity to one entrance of a resource to help users grasp the actual meaning of this entity. For the reason that traditional entity linking methods cannot acquire high performance in both accuracy and efficiency, we propose a novel entity linking algorithm. This algorithm is mainly divided into three steps. It first generates candidate entities for each mention in documents via heuristic-based rule. Then we leverage the relationship between entities in the knowledge base and use them to construct a semantic entity graph to connect all the related candidate entities. Finally we give a score to measure the possibility of one entity to be an entrance for one mention and choose the one with the highest score as the best assignment. Experimental results show that our entity linking algorithm performs well in both accuracy and efficiency.

Keywords: Collective entity linking · Entity disambiguation · PageRank Semantic entity graph

1 Introduction

Nowadays, Internet has crossed almost all the aspects of modern life and generated a large amount of text source. However, it is difficult to obtain the intention of users in search engine due to the multiple expressions of entity in natural language. One example is the full name versus aliases or acronyms. Given a sentence "Steve Jobs is a co-founder of Apple", "Apple" may refer to a kind of fruit, an American multinational technology company, or the name of one film. To require entity's actual meaning, the most convenient way is to map name mention to its corresponding entity in the knowledge base. Therefore "Apple" in the aforementioned sentence should link to the entity "Apple_Inc." in Wikipedia. This approach is called entity linking. There are multiple knowledge bases available, such as WordNet, YAGO2, Freebase, DBpedia, etc. Since DBpedia is the structural form of Wikipedia and due to the extensive use of Wikipedia, we put DBpedia as the fundamental knowledge base in our algorithm. It contains rather than 11 million entity nodes and more than 580 million records that describe the relationship between nodes (or entities) in English version. It has a high coverage of named entities and satisfied the demand of our task.

© Springer Nature Singapore Pte Ltd. 2017
J. Li et al. (Eds.): CCKS 2017, CCIS 784, pp. 66–79, 2017.
https://doi.org/10.1007/978-981-10-7359-5_8

In this paper, we design an entity linking algorithm and it contains two modules: candidate entity generation and candidate entity selection over DBpedia. In candidate entity generation module, we leverage search rule and prior probability to gain candidate entity sets. In candidate entity selection module, we construct a graph by recording the links between entities, where one node represents one candidate entity or one bridged entity (the entities that can form a path to connect two candidate entities in DBpedia). Edge between two nodes denotes that there exists the relation between two entities corresponding to the two nodes. Through calculating the PageRank value of each node, we treat the candidate entity with the highest PageRank value as the best assignment for one name mention. Finally, we use TAC-KBP 2014, AIDA-EE and AIDA CoNLL-YAGO as testing corpora, and experimental result indicates that our algorithm achieves both high accuracy and low running time. The innovations of our method are as follows:

1. We generate candidate entities based on users' search patterns and the probability of entities' appearance in DBpedia.
2. We construct an entity graph by making full use of the relationships between entities in DBpedia.
3. Since there is a high cohesion between entities in one document, we utilize the PageRank value of one node to represent the cohesion of the entity in the constructed graph.

2   Relate Work

As far as we know, an entity can be mentioned in the context in many forms such as full name, aliases, abbreviation and so on [1]. That means a name mention may have many possible candidate entities which refer to this name mention in several specific contexts.

Many approaches related to entity linking have been put forward. In general, there are mainly two models to the problem. One is independent entity model, which is based on the hypothesis that name mentions in the context are independent on each other. Cucerzan et al. [2] constructed a system for the recognition and disambiguation of the named entities. Through comparing with the contextual information of one name mention and its one possible candidate entity in the knowledge base, one candidate entity is selected by the category tags associated with this name mention. Hanet et al. [3] leveraged multiple types of entity knowledge such as popularity knowledge, name knowledge and context knowledge. Zuo et al. [4] put forward a method called bagging for entity linking through just operating on a textual range of relevant terms. It makes the decision by voting from multiple simple classifiers. Bunescu et al. [5] utilize s a series of information including the context of name mentions, Wikipedia article of candidate entities and Wikipedia category and so on to calculate the similarity between candidate entities and name mentions. Daiber et al. [6] used the combination of features and train these data via linear regression.

The other one is collective entity linking model that considers name mentions in the context together to help determine the best entity associated with on name mention [7]. Alhelbawy et al. [8] treated named entities as hidden variables and use HMMs to resolve the name mention disambiguation problem. In general, the method based on graph is a more common approach in collective entity linking model, which constructs the graph to express the high cohesion between entities in one document. Hoffart et al. [9] constructed a weighted graph according to the prior probabilities of entities, the relationships between name mentions and candidate entities. Then the entities with strong probabilities are selected as mapping entities for name mentions. Hachey et al. [10] adopted an unsupervised approach to exploit the knowledge from Wikipedia to form a linked graph. Ceccarelli et al. [11] generated a learning to rank formulation for discovering the most suitable entities. Usbeck et al. [12] combined HITS algorithm with label expansion and string similarity measures, which achieved a great performance on MSNBC dataset. Florian et al. [13] collected local features and global features for each entity, and then adopted the CRF-based model to determine the mapping entity.

In the aforementioned methods, the ones like these based on independent entity model cannot acquire high accuracy due to no considering the relation between name mentions. By contrast, the ones with some complex system designs may influence the running efficiency. To balance accuracy and efficiency, we propose a novel approach based on collective entity linking model. We first leverage the features from DBpedia [14] to obtain the candidate entities of name mentions in document. Then we construct a graph via the relation between candidate entities in knowledge base. Finally, through calculating the scores of candidate entities, we select the top one as the mapping entity of name mention in DBpedia. The details are shown as follows.

3   Preliminary

In Text Analysis Conference, entity linking is regarded as one of the subtasks in the knowledge base population track. Before introducing our method, it is necessary to clarify several concepts and files about Wikipedia that are used in our method.

DBpedia. DBpedia is a structured multi-language corpus extracted from Wikipedia by DBpedia Community. It uses RDF (Resource Description Framework) to describe the relation between entities. The covered field of DBpedia's English version is very comprehensive. It records more than 11 million entities, including person, film, book and so on, we use the English version of DBpedia as knowledge base in our task.

Bridging Entity (called bridging node in our entity graph). In our method, we need to detect the paths between candidate entities through the relation among entities in DBpedia. In general, there is no direct connection between most of entities in DBpedia. Instead, we need to use some bridging entities to connect candidate entities. For example, given two candidate entities "Michael Jordan" and "Yao Ming", there is no direct links between them, but the two candidate entities can be connected by the bridging entity "National Basketball Association", forming a path "Michael Jordan→National Basketball Association→Yao Ming". Therefore in the graph, there exist two types of entities, the candidate entities and the bridging entities.

Candidate Entity. It is well-known that one name mention may have several meanings. The entities in the knowledge base that correspond to these meanings are called candidate entities for this name mention.

Mapping Entity. In candidate entity selection model, we select a candidate entity that has maximum likelihood with name mention. This kind of candidate entity is called mapping entity.

Labels_en.nt. It is the corpus that contains all titles of entity in Wikipedia. This file is used to generate the candidate entities that are similar with name mention. These candidate entities will be included in candidate set.

Redirects_en.nt. It records the redirect pages in Wikipedia. In the redirect page, multiple name mentions refer to one entity. For example, when "PRC" or "People's Republic of China" mentioned in one document, they both point to the entity "China" in Wikipedia to denote that "PRC" and "People's Republic of China" is fulltime and abbreviation of "China".

Disambiguations_en.nt. It stores the content of disambiguation pages. For certain entity that exists multiple meanings, it will be listed in Disambiguations_en.nt. Most of entities in disambiguation pages is equipped with "(disambiguation)" behind entity name. For instance, when we retrieve "Michael Jordan" in Disambiguations_en.nt, we can obtain "Michael Jordan", "Michael Jordan (footballer)", "Michael Jordan (mycologist)" and so on. Since there are mostly ambiguous named mentions in text, the disambiguation pages offer candidate entities to select.

Pagelinks_en.nt. It is the corpus that is created from the internal links between Wikipedia articles and records the relationship between entities. When we construct the entity graph, we need to utilize the link relationships between entities as edges in the graph. For instance, a word "with its seat of government in the capital city of Beijing" appeared in entity page "China". We find that there is relation between "China" and "Beijing". That is to say, there is an edge ("China", "Beijing") in the entity graph.

4 Method Description

Our entity linking method is mainly divided into two modules: candidate entity generation and candidate entity selection. In candidate entity generation module, for each name mention in one document, we retrieve a candidate entity set which contains several possible entities that one name mention may refer to in DBpedia. In candidate entity selection module, we construct an entity graph, which of the nodes represent candidate entities and the edges represent the relationships between entities. Via this graph, PageRank values of each candidate entity are calculated, and the candidate entity with the top value is treated as the best assignment. Below are the details about our method.

4.1 Candidate Entity Generation

For one name mention $m \in M$ in the document, the goal of this module is to generate a candidate entity set E_m from DBpedia. Each entity in E_m is one possible candidate that m can be referred to. The general method for this goal is to use Edit Distance, which counts the operation times from changing one name mention to its one candidate entity in DBpedia. However, this method has some major shortcomings. For instance, it may generate a large scale of candidate entities for each name mention. Therefore, this method is not practical. Besides, the Wikipedia API can be retrieved by fuzzy query. When we input a string in query, it will return a series of entities relevant to the string. However, it is online candidate generation approach so we need to connect network when we request data through this method. To solve the drawbacks of these methods, we propose two approaches to generate candidate set.

Candidate Entity Generation Based on Search Rule

Name mention in text is usually a nonstandard representation form of entity, such as abbreviation, nickname and even spelling mistake [15]. In our candidate entity generation module, we generate candidate sets using the following four steps:

1. If one name mention can be retrieved in Labels, it will be added to candidate set.
2. If one name mention can be retrieved in Redirects, the redirect entity set corresponding to this name mention will be added to candidate set.
3. If one name mention can be retrieved in Disambiguation, the disambiguations entity set corresponding to this name mention will be added to candidate set.
4. If the redirect entity set is not empty, the entities in Labels (if exist) and the entities in Disambiguations (if exist) corresponding to redirect entities will be added to candidate set.

The approach mentioned above accords with our normal search rule. When we want to retrieve a word, what we enter may be a nonstandard expression. As the same, if one entity is the same with what we enter, this entity will be returned directly. Otherwise it will lead to disambiguation page. Sometimes when what we search is an alias of a person or a place, we will be guided to redirect page.

Candidate Entity Generation Based on Prior Probability

As far as we know, anchor text is a common element in entity pages, which is equal to hyperlink of an entity. We can jump from one Wikipedia page to another page by anchor text. In general, anchor text is the nickname of mapping entity. For example, for the anchor text "Big_Apple", "New_York" is its referent entity. They are expressed using the format [New_York|Big_Apple] in MediaWiki.

We collect all entity pages in Wikipedia and count the times of entity corresponding to anchor text in entity pages. Then we get the prior probability $P(e_i|m)$ using the following formulas.

$$P(e_i|m) = \frac{P(m, e_i)}{P(m)} \tag{1}$$

$P(m, e_i)$ denotes the times of the concurrence of anchor text m and its corresponding entity e_i appearing in Wikipedia article. $P(m)$ denotes the times of the anchor text m appearing in entity pages.

In our method, if the value of $P(e|m)$ is higher, the probability of the mapping entity e appearing in DBpedia is larger on the condition that the name mention m is already appearing in document. For example, given that $P(\text{"}China\text{"}|\text{"}China\text{"})$ is 0.83 and $P(\text{"}China(song)\text{"}|\text{"}China\text{"})$ is 0.01, which mean that if "China" appears in document, the possibility of the mapping entity called "China" (a country name) in DBpedia is very high. By contrast, the possibility of the mapping entity called "China" (a song) is correspondingly very low. Based on the above hypothesis, we choose the top 10 entities through prior possibility rank and add them to candidate set.

In conclusion, when candidate set is empty, we generate candidate set by Search Rule. Otherwise, we use Prior Probability to expand candidate set. Since the approach of Prior Probability used to obtain candidate set only needs to calculate probability, it consumes less time than Search Rule method.

4.2   Candidate Entity Selection

In this module, we construct an entity graph through utilizing the relationships between entities. In general, one document expresses one topic or several similar topics [16]. Based on the topic consistency, we deem that the probability that one candidate entity is a mapping entity for one name mention depends on the relationships between this candidate entity node and the other entity nodes in the graph constructed by the following algorithm.

Graph Construction

DBpedia is a large-scale semantic knowledge graph and records the relationships between entities in the form of triple. Based on the correlations of entities in DBpedia, we construct entity graph by expanding candidate entities to import some bridging entities in DBpedia.

Suppose that there exists one name mention set $M = \{m_1, m_2, \dots, m_i\}$ and one candidate set $E_{m_j} = \{e_{i1}, e_{i2}, \dots, e_{ij}\}$ corresponding to the name mention m_i. Given any two candidate entities e_{ij} and e_{mn}, we leverage DFS (Depth-First-Search) algorithm to retrieve the paths between e_{ij} and e_{mn}. Following this step, many paths that can connect two candidate entities can be found. Different from candidate entities, the other entities in these paths are called bridging entities. With these bridging entities, we construct an entity graph. In other words, the entity graph we design can be treated as a subgraph of semantic knowledge graph extracted from Wikipedia and is just applicable to current document. The graph construction algorithm is shown as follows.

ALGORITHM 1. Graph Construction

Input: Name mention set M , and candidate sets $E_m = \{E_{m_i} \mid m_i \in M\}$.

Output: $G(V, E)$

1. Initial setting V, E = NULL

    /*begin to construct graph*/
2. **for** m_i in M :
3.   **for** m_j in M :
4.   **if** E_{m_i} is not empty and E_{m_j} is not empty:
5.   choose e_{ik} from E_{m_i} and e_{jl} from E_{m_j}

              /* detect the paths between e_{ik} and e_{jl} from DBpedia*/
6.   path set *Paths* = Depth-First Search(e_{ik} , e_{jl} , *maxlength*)

              /* *maxlength* defines the max length of path*/
7.   **for** *node* in *Paths* :
8. V .add(*node*)
9.   **end for**
10.   **for** *edge* in *Paths* :
11. E .add(*edge*)
12.   **end for**
13.   **end if**
14.   **end for**
15. **end for**

One example graph is shown in Fig. 1.

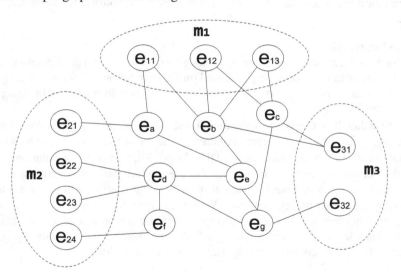

Fig. 1. Entity graph

As shown in Fig. 1, suppose $\{m_1, m_2, m_3\}$ is the name mention set in the document D. There are several candidate entities for per name mention in the graph. For instance, e_{11}, e_{12}, e_{13} denote the candidate entities for m_1. The other nodes such as $e_a \ldots e_g$ are bridging entities, which connect candidate entities by paths. For example, when we retrieve the paths between e_{13} and e_{31} according the relationships in DBpedia, we finally find two paths (e_{13}, e_b, e_{31}) and (e_{13}, e_c, e_{31}).

Candidate Entity Ranking

Since one document depicts one topic or similar topics, we deem that if one candidate entity has more relationships with the other candidate entities, there should be many paths to connect this candidate entity with the other entity nodes in our graph. That indicates the probability of the fact that this candidate entity is the corresponding entity for one name mention is higher [17]. Based on this fact, we calculate the PageRank value of each candidate entity to determinate whether this candidate entity is one mapping entity for one name mention or not.

The reason to use PageRank value to measure the possibility of assigning one candidate entity to one entity mention is that PageRank value of the node depends on the number of edges between this candidate entity and the other entity nodes in the graph. This situation accords with our application. Therefore it is a good choice to calculate PageRank values of candidate entities to determine whether this candidate entity is one mapping entity or not.

PageRank value of one node can be calculated by

$$PR_i = \sum_{j \in E_i} \frac{PR_j}{N_j} \tag{2}$$

PR_i denotes the PageRank value of one node i, E_i represents the nodes that are linked to page i. N_j represents the number of node j that links to the other nodes. In our entity graph, we calculate the PageRank value of each node and choose the candidate entity which owns the highest value for name mention m_i as the mapping entity of m_i.

5 Evaluation

5.1 Dataset Description

There are three datasets that are used to evaluate our method, including TAC-KBP 2014, AIDA-EE, and AIDA CoNLL-YAGO. TAC-KBP 2014 mainly focuses on long text. KBP (Knowledge Base Population) is an evaluating task of NIST (National Institute of Standards and Technology). It aims at unstructured text construction and knowledge base extension technology [18]. TAC-KBP 2014 contains 466 documents and 11670 name mentions marked in text. Each name mention corresponds to a mapping entity in Wikipedia. AIDA-EE contains 300 documents with 9976 entity names linked to Wikipedia. The documents are taken from the APW part of the GIGAWORD5 dataset with 150 documents from 2010-10-01 and 150 documents from 2010-11-01, most of the name mention in which with high ambiguity. AIDA CoNLL-YAGO [19] contains assignments of entities to the mentions of named entities annotated for the original CoNLL 2003 entity recognition task.

5.2   Evaluation Criteria

NIL entity problem is worthy to be considered in entity linking. For instance, the referent entity of name mention may not be contained in the given knowledge base. Under this circumstance, this name mention should be linked to a NIL entity. In our method, we only consider non-NIL entity mentions and use standard precision, recall and F-measure to evaluate entity linking performance [20].

$$Precision = \frac{N_c - NIL_p}{N - NIL_p} \tag{3}$$

$$Recall = \frac{N_c - NIL_p}{N - NIL_r} \tag{4}$$

$$F_1 = \frac{2 * Precision * Recall}{Precision + Recall} \tag{5}$$

where N denotes the number of all the name mentions. N_c represents the number of the name mentions to be predicted correctly. NIL_p denotes the number of NIL entity to be predict. NIL_r shows the number of actual NIL entity.

5.3   Experimental Results

In Table 1, we obtain the performances of different methods in candidate entity generation module in TAC-KBP 2014. They include Edit Distance, Wikipedia API and our proposed method. In Table 1, Coverage rate denotes the ratio of the corresponding entity included by candidate set. Average size denotes the average number of candidate set.

Table 1. Performances of candidate entity generation by different methods

Method	Coverage rate	Average size
Edit Distance (6)	64.77%	287
Wikipedia API (5)	86.84%	5
Wikipedia API (10)	89.31%	10
Search Rule	92.44%	14.9
Prior Probability	91.57%	6.89

From Table 1, we can see that Edit Distance is not a good approach, due to its low coverage and large candidate entity number, which brings great challenge to select candidate entity. Wikipedia API may be a workable method. When the number of entities returned is set as 10, average rate of correct mapping entity reaches 89.31%. However, Wikipedia API is an online request application and needs to consume long time to return results. Compared with the other methods, we find that two approaches we proposed perform well in both coverage rate and average number. Table 2 tells the performance of candidate entity generation by different methods in different datasets.

Table 2. Performances of candidate entity generation by different methods in different datasets

Method	TAC-KBP 2014	AIDA-EE	CoNLL-YAGO
Edit Distance (6)	65.53%	62.36%	61.17%
Search Rule	92.54%	71.45%	74.72%
Prior Probability	90.81%	79.83%	83.96%

Table 2 compares the three methods' performances of candidate entity generation in different datasets. As shown in Table 2, we find Prior Probability method has works slightly better than Search Rule method. To further analyze the contrition of Prior Probability method and Search Rule method, we test the overall performance of these two approaches in TAC-KBP 2014 in the case that there are no more than 2 paths between two candidate entities in Table 3. As we can see, the performance slightly degraded in Prior Probability method. However, the Search Rule method needs more running time than Prior Probability method, because there exists a larger scale of graph and each name mention has more candidate entities to rank in Search Rule method. So if we want to reduce running time and obtain higher accuracy, Prior Probability method is a better choice.

Table 3. System performances by different candidate entity generation methods

Method	Precision	Recall	F1
Search Rule	85.45%	85.00%	85.22%
Prior Probability	84.36%	83.92%	84.14%

Since TAC-KBP 2014, AIDA-EE and AIDA CoNLL-YAGO are generally used datasets for entity linking. We adopt the default candidate entity generation approach of Prior Probability, and the results of using these datasets are displayed in Table 4.

Table 4. System performances in different datasets via default candidate entity generation approach

Test set	Precision	Recall	F1	Runtime per doc (s)
TAC-KBP 2014	84.36%	83.92%	84.14%	21.20
AIDA-EE	69.21%	73.55%	71.32%	337.17
CoNLL-YAGO	71.07%	71.07%	71.07%	82.29

As shown in Table 4, the performances of our method on TAC-KBP 2014 is well. However its performances tested on the other two datasets are not very well. The main reason is that there are many new entities corresponding to name mentions appeared in DBpedia, which have few relationships with the entities in DBpedia. AIDA-EE and CoNLL-YAGO both have lots of new entities, which brings in great difficulties to link entities in DBpedia to name mentions appearing in these two datasets.

As the size of text grows larger, the cohesion of entities which are far apart in one text deteriorates. Therefore, we try to set different window values to divide a document to improve the performances on the AIDA-EE and CoNLL-YAGO and reduce running time. When the window value is set 5, it means that we divide the name mentions in one document into the nearest 5 name mention per group. As shown in Fig. 2, the precision gets the highest precision in the three datasets when the window value is set 10. Additionally, the smaller the window is set, the less the running time takes.

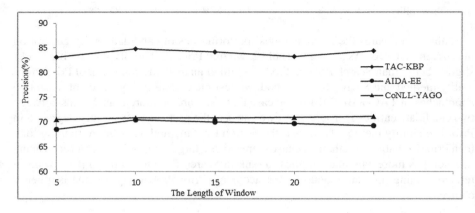

Fig. 2. Precision results when window is set in different values

To improve our system's performance, we apply the greedy search method to select the highest score as the best assignment. In the greedy search method, we choose the maximum likelihood PageRank value of the node and update the graph every time. As seen in Table 5, the greedy search method works a little better than Table 4. While the greedy search make a little progress, we don't adopt it because it consumes long running time.

Table 5. System performances in different datasets via greedy search

Test set	Precision	Recall	F1
TAC-KBP 2014	85.87%	85.69%	85.51%
AIDA-EE	70.86%	75.06%	72.89%
CoNLL-YAGO	72.01%	72.01%	72.01%

To get rid of the coverage of the dataset's influence, we assume every candidate entities set include the name mention's corresponding entity in every text. In the process of generating candidate set, we add the corresponding entity in the candidate set.

Table 6. System performances in different datasets on the assumption that right entity are included by each candidate entity set

Test set	Precision	Recall	F1
TAC-KBP 2014	89.33%	89.15%	88.96%
AIDA-EE	87.18%	87.05%	86.94%
CoNLL-YAGO	78.47%	78.47%	78.47%

Table 6 reveals that our method make great achievement in candidate entity selection after the coverage rate of corresponding entity becomes 100%.

Based on the previous situation, we use the TAC-KBP 2014 as basic testing corpus and compare our method with some state-of-the-art approaches on entity linking. The baselines include independent entity model such as Spotlight [11] and collective entity model such as AGDISTIS [12], IBM [13]. Among these baseline methods, Spotlight adopts the dependence among co-occurence. And AGIDISTIS is the same method as our graph-based method on entity linking with DBpedia, but it applies HITS to entity linking. IBM uses language independent probabilistic disambiguation model for the TAC-KBP dataset. They are state-of-the-art approaches on entity linking. The results are presented as follows.

By observing Table 7, we can see that our method performs the best in both precision and recall. In these baselines, some of them adopt the dependence among co-occurrence features, like Spotlight, which results in generating too many spurious candidate entities and reduces accuracy. Some of them are based on graph, which adopts too many extra properties to excessively express the relationships between entities and increases running time. Considering theses weaknesses, our method utilizes paths to connect related candidate entities to form a graph, and adopts PageRank value to measure the cohesion of one candidate entity with the other entities in the graph, which can better grasp the cohesion between candidate entities to acquire high accuracy. In more detailed analysis, our method uses Prior Probability method and Search Rule method to generate candidate entity, which performs well in both coverage rate and average number. And the method we adopted in entity linking attaches great importance to the high cohension of the document and the semantic relationships between entities in knowledge base. Experimental results also support our method.

Table 7. System performance of different methods

System	Precision	Recall	F1
AGDISTIS	77.70%	55.60%	64.28%
IBM	80.60%	77.70%	79.10%
Spotlight	83.30%	79.60%	81.40%
Our system	85.45%	85.00%	85.22%

6   Conclusion and Future Work

In the light of large scale ambiguous texts on Internet, we design a new method for entity linking using DBpedia. Our method is divided into two parts. One is candidate entity generation and another one is candidate entity selection. For the first part, we propose two combined approaches, one based on search rule and the other one based on prior probability. Then candidate entities are expanded according to the relationships between entities in DBpedia and an entity graph about the current document is constructed based on those relationships. This graph is just a subgraph of a semantic knowledge base (in our method it is DBpedia). Finally we calculate PageRank value of each node in the entity graph and select the candidate entity with the largest value as the best assignment for one name mention. We use TAC-KBP 2014, AIDA-EE, AIDA CoNLL-YAGO to evaluate our method and experimental results are relatively better compared with some state-of-the-art approaches.

Our method still has some weaknesses. In the future work, we intend to improve its performance in the following points. Firstly, despite that the efficiency of our method is already high, there also exists space to be optimized. Secondly, the entity graph constructed by this paper contains lots of nodes and edges. They need to be calculated through a great quantity of iterations in PageRank algorithm. Therefore we hope to adopt some other algorithms such as Monte Carlo random walk and mountain climbing to optimize calculating process. Finally, the final accuracy of our method is also not ideal. We have to improve the accuracy by considering more features in entity's page.

Acknowledgements. The research in this paper is supported by National Natural Science Foundation of China (No. 61300114, 61572151), CCF-Tencent Open Fund (No. IAGR20160109), HIT-Tecncent (No. AGR201601).

References

1. Shen, W., Wang, J., Han, J.: Entity linking with a knowledge base: issues, techniques, and solutions. IEEE Trans. Knowl. Data Eng. **27**(2), 443–460 (2015)
2. Cucerzan, S.: Large-scale named entity disambiguation based on wikipedia data. In: EMNLP-CoNLL, vol. 7, pp. 708–716 (2007)
3. Han, X., Sun, L.: A generative entity-mention model for linking entities with knowledge base. In: Proceedings of the 49th Annual Meeting of the Association for Computational Linguistics: Human Language Technologies, vol. 1, pp. 945–954. Association for Computational Linguistics (2011)
4. Zuo, Z., Kasneci, G., Gruetze, T., et al.: BEL: Bagging for Entity Linking. In: COLING, pp. 2075–2086 (2014)
5. Bunescu, R.C., Pasca, M.: Using encyclopedic knowledge for named entity disambiguation. In: EACL, vol. 6, pp. 9–16 (2006)
6. Daiber, J., Jakob, M., Hokamp, C., et al.: Improving efficiency and accuracy in multilingual entity extraction. In: Proceedings of the 9th International Conference on Semantic Systems, pp. 121–124. ACM (2013)
7. Schäfer, B., Bizer, C.: Exploiting DBpedia for graph-based entity linking to wikipedia. FakultätfürWirtschaftsinformatik und Wirtschaftsmathematik (2014)
8. Alhelbawy, A., Gaizauskas, R.: Named entity disambiguation using HMMs. In: 2013 IEEE/WIC/ACM International Joint Conferences on Web Intelligence (WI) and Intelligent Agent Technologies (IAT), vol. 3, pp. 159–162. IEEE (2013)
9. Hoffart, J., Yosef, M.A., Bordino, I., et al.: Robust disambiguation of named entities in text. In: Proceedings of the Conference on Empirical Methods in Natural Language Processing, pp. 782–792. Association for Computational Linguistics (2011)
10. Hachey, B., Radford, W., Curran, J.R.: Graph-based named entity linking with wikipedia. In: Bouguettaya, A., Hauswirth, M., Liu, L. (eds.) WISE 2011. LNCS, vol. 6997, pp. 213–226. Springer, Heidelberg (2011). https://doi.org/10.1007/978-3-642-24434-6_16
11. Ceccarelli, D., Lucchese, C., Orlando, S., et al.: Learning relatedness measures for entity linking. In: Proceedings of the 22nd ACM International Conference on Information and Knowledge Management, pp. 139–148. ACM (2013)

12. Usbeck, R., Ngomo, A.C.N., Röder, M., et al.: AGDISTIS-agnostic disambiguation of named entities using Linked Open Data. In: ECAI, pp. 1113–1114 (2014)
13. Sil, A., Florian, R.: The IBM systems for trilingual entity discovery and linking at TAC 2015. In: Proceedings of Text Analysis Conference (TAC 2015) (2015)
14. Kalloubi, F., Nfaoui, E.H., El Beqqali, O.: Graph based tweet entity linking using DBpedia. In: 2014 IEEE/ACS 11th International Conference on Computer Systems and Applications (AICCSA), pp. 501–506. IEEE (2014)
15. Bhattacharya, I., Getoor, L.: Online collective entity resolution. In: Proceedings of the National Conference on Artificial Intelligence, Menlo Park, CA, vol. 22, no. 2, p. 1606. MIT Press, Cambridge, AAAI Press, London (2007)
16. Ratinov, L., Roth, D., Downey, D., et al.: Local and global algorithms for disambiguation to wikipedia. In: Proceedings of the 49th Annual Meeting of the Association for Computational Linguistics: Human Language Technologies, vol. 1, pp. 1375–1384. Association for Computational Linguistics (2011)
17. Blanco, R., Ottaviano, G., Meij, E.: Fast and space-efficient entity linking for queries. In: Proceedings of the Eighth ACM International Conference on Web Search and Data Mining, pp. 179–188. ACM (2015)
18. Nguyen, D.B., Hoffart, J., Theobald, M., et al.: AIDA-light: high-throughput named-entity disambiguation. In: LDOW (2014)
19. Luo, G., Huang, X., Lin, C.Y., et al.: Joint named entity recognition and disambiguation. In: Proceedings of EMNLP (2015)
20. Ji, H., Nothman, J., Hachey, B.: Overview of TAC-KBP2014 entity discovery and linking tasks. In: Proceedings of Text Analysis Conference (TAC 2014) (2014)

A CWTM Model of Topic Extraction for Short Text

Yunlan Diao[✉], Yajun Du, Pan Xiao, and Jia Liu

School of Computer and Software Engineering, Xihua University, Chengdu 610039, China
240582045@qq.com

Abstract. The topic model is designed to find potential topics from the massive micro-blog data. On the one hand, the extraction of potential topics contributes to the next analysis. On the other hand, because of the particularity of the data, we can not deal with it directly with the traditional topic model algorithm. In the field of data mining, although the traditional text topic mining has been widely studied, a short text like micro-blog has the distinctive characteristics of network languages and emerging novel words. Owning to the short message, the sparsity of data and incomplete description, the micro-blog can not be obtained efficiently. In this paper, we propose a simple, fast, and effective topic model for short texts, named couple-word topic model (CWTM). Based on Dirichlet Multinomial Mixture (DMM) model, it can leverage couple word co-occurrence to help distill better topics over short texts instead of the traditional word co-occurrence way. The method can alleviate the data sparseness problems, improve the performance of the model and adopt the Gibbs sampling algorithm to derive parameters. Through extensive experiments on two real-world short text collections, we find that CWTM achieves comparable or better topic representations than traditional topic model.

Keywords: Topic model · Short texts · Couple word

1 Introduction

Social media is a special kind of text, severely restricted (within 140 words), which lacks rich contextual information and semantic information. Therefore, the traditional topic model can not analyze the theme characteristics of social media well. However, topic model has played a significant role in dealing with traditional text mining and can solve the polysemy problems, which proves that the topic model has more advantages over other approaches in the analysis of the theme characteristics of the text, especially in language intensive Chinese text. Ordinary users of the Internet generate a vast amount of short text data through social media every day, and their information are unmatched with traditional media. Therefore, the research on how to use topic models to extract

Y. Du—This work is supported by the National Nature Science Foundation (Grant No. 61472329 and 61532009), the Key Natural Science Foundation of Xihua University (Z1412620) and the Innovation Fund of Postgraduate, Xihua University.

© Springer Nature Singapore Pte Ltd. 2017
J. Li et al. (Eds.): CCKS 2017, CCIS 784, pp. 80–91, 2017.
https://doi.org/10.1007/978-981-10-7359-5_9

effective topic features from the short text of massive social media has become a hot topic in the international academic community.

Short texts, e.g., Micro-blog, news headlines, and reviews data, have become a fashionable form of information on the Internet. Micro-blog is one of the currently mainstream social networking platform, which has become an important means for users to publish and access to real-time information, as for many data analysis applications involving extracting the underlying theme from these short texts in micro-blog. On the one hand, the extraction of potential topics contributes to the next analysis, such as user interest profiling [1], topic detection [2], comments summarization [6], text classification [10] and content characterizing [14]. On the other hand, because of the particularity of the data, we can not deal with it directly with the traditional topic model algorithm. Topic model mining has been proved to be a very effective means for text topic mining. With the development of timely communication, it is becoming more and more significant to derive the theme of short text features.

The purpose of micro-blog topic modeling is to excavate interesting topics and other user information from huge amounts of information. Micro-blog has the characteristics of short message text, fast updating of information and large amount of data, and of existing the phenomenon of multi-word meaning. For example, "dollar", "$", "$$", "fee", "charges" have the same meaning. Limited in the length of the text, it is difficult to directly extract the information from the passage of the data, leading that the theme traditional model methods can not effectively dig out information users really interested in. The object of our study is Chinese micro-blog. Micro-blog in Chinese is a little different from in English, and each word in English is a complete semantic word while in Chinese is different from where a single word can not form a complete semantic word, which requires us to handle the text segmentation into a separate full sense of the word. Unlike long documents, the short text usually contains only one topic. This may seem like a good idea, but traditional topic model algorithms assume that a document contains multiple topics, which can cause a lot of trouble in modeling analysis.

In the past more than ten years, as an important branch of data mining, the topic model has made great development and progress, and has played an significant role in many fields, such as media, business, biology, and medicine and so on. For example, Fabrice et al. implemented a location based video retrieval system using LSA, Keith et al. had applied LDA [13] to mass image community mining, Phan et al. completed a cluster of short texts based on the concept of topic model. Driven by the unremitting efforts of scholars and social needs, the concept of topic model has been accepted by more and more scholars and research institutions. Therefore more and more scholars have devoted themselves to the research of topic model, and have turned these research results into practical products, bringing convenience to our daily work and life.

Although many scholars both at home and abroad have done a lot of researches on the topic model algorithm and put forward some methods of modeling the theme of the short text, these methods are still unable to fully meet the needs of practical work, the existing problems mainly in three aspects. First, the ability to model short text is inadequate. The short texts can be divided into two parts: general short passage and key word essay represented by the title of the paper and the headline of the news. In a keyword essay, each word has the same weight, so the frequency of the word appearing in the text is not significant. However, the general topic model is based on the frequency of occurrence of words in the text, without the optimization of the feature of the key words. Second, the production of low-quality topics cannot be avoided. These classical algorithms extract topics according to the user specifying the number of topics, which can not guarantee that each topic modeling result has obvious characteristics and high quality data in colleagues concentrated on theme having a certain weight, resulting in the modeling results not only giving the daily work convenience, but the fuzzy theme feature giving users a trouble. Finally, it is easy to cause overlapping between subjects. Although the ultimate purpose of topic modeling algorithm is also produced obvious characteristics of independent collection of topics, in the existing model of the theme, the situation of two themes combined into the same theme in the modeling results is unavoidable.

Topic extraction method of micro-blog short text proposed in this paper expands the semantic information of short text by extracting couple words, so as to alleviate the data sparseness problem of short text and to improve the effect of topic features mining. We assume that the document is a mixture of topics and each micro-blog short text consists of multiple couple word which is independent of each other. What's more, the word pairs are unordered and belong to the same topic.

The main contributions of this paper are summarized as follows:

1. We develop a new simple, fast, and effective model for topic extraction over short texts. In this case, acquiring couple word co-occurrence to model instead of the traditional word co-occurrence way. The method can alleviate the data sparseness problem and improve the performance of the model.
2. On two real-world short test collection, we evaluate the proposed CWTM against a few state-of-the-art alternatives for Micro-blog text. Experimental results demonstrate our approach can discover more prominent and coherent topics, and significantly outperform baseline methods.

The remainder of this paper is organized as follows. In Sect. 2, we introduce related work on topic models for short text. Section 3 describes our models for short text topic modeling. Section 4 introduces the experimental methodology and shows the results obtained. Finally, conclusions and future work are made in the last section.

2   Related Work

In recent years, the research on feature extraction of short text topic has received extensive attention and been widely studied, and various models and methods have sprung up. In this section, we review recent advances on learning better topic representations on short texts. At present, more and more researches have been done on the topic extraction model in English micro-blog environment, but the topic extraction for Chinese micro-blog is still in its initial stage.

Topic models have been proposed to uncover the latent semantic structure from text corpus. The effort of mining semantic structure in a text collection can be dated from latent semantic analysis (LSA) [12], which utilizes the singular value decomposition of the document-term matrix to reveal the major associative words patterns. Many more complicated variants and extensions of LDA and Probabilistic latent semantic analysis (PLSA) [5] which are the conventional topic modeling techniques have been proposed, such as the supervised topic model [4], authortopic model [8], and Bayesian nonparametric topic model [14]. In PLSA, a document is presented as a mixture of topics, while a topic is a probability distribution over words. PLSA improves LSA with a sounder probabilistic model based on a mixture decomposition derived from a latent class model. Extending PLSA, LDA adds Dirichlet priors on topic distributions, resulting in a more complete generative model. Due to its nice generalization ability and extensibility, LDA achieves huge success in text mining domain. These models and their variants have been studied extensively for various tasks in information retrieval and text mining. However, both of them utilize word co-occurrences as structure priors for topic-word distribution, rather than directly modeling their generation process.

Recently, many efforts have been spent towards intensifying the word co-occurrence information from the collection of short texts being modeled. Biterm Topic Modeling (BTM) [3], proposed by Yan et al., learns topics by directly modeling the generation of biterms, i.e., pairs of words that co-occur in the same document. The model aggregates all corpus biterms in a big pseudo-document that is used to infer the topic distribution, overcoming the sparsity problem at a document level. BTM can well capture the topics within short texts as it explicitly models the word co-occurrence patterns and uses the aggregated patterns in the whole corpus. Since the inference is done over one pseudo-document, the model has a single topic distribution for the entire corpus, instead of one distribution per document. The experimental results show that BTM outperforms existing state-of-the-art alternatives in terms of effectiveness and efficiency. Inspired by the aforementioned aggregation strategies, Quan et al. propose a self-aggregation based topic model (SATM) [16] for short texts. SATM assumes that shares the same topic proportion of the latter and each short text is a segment of a long pseudo-document. In these models, each document is represented as a multinomial distribution over topics and each topic is represented as a multinomial distribution over words.

In recent years, topic models have been utilized for social media content analysis in various tasks with the emergence of social media. Statistical techniques (e.g., Gibbs sampling) are then employed to identify the underlying topic distribution of each document as well as word distribution of each topic, based on the high-order word co-occurrence patterns. However, due to the lack of specific topic models for short texts, some researchers directly applied conventional (or slightly modified) topic models for analysis [7, 10, 23]. Some others tried to aggregate short texts into lengthy pseudo-documents based on some additional information, and then train conventional topic models [18, 19]. Hong et al. [22] made a comprehensive empirical study of topic modeling in Twitter, and suggested that new topic models for short texts are in demand.

In this paper, we propose a novel topic model for short texts, named CWTM based on DMM, to leverage couple word co-occurrence to help distill better topics over short texts instead of the traditional word co-occurrence way. Although we do not introduce new topic models to address the issues of short text modeling especially in microblogging environments in this paper, our work sheds some light on how research on topic models can be conducted for short text scenarios.

3 A Novel Topic Modeling

Conventional topic models learn topics based on document level word co-occurrence patterns, whose effectiveness will be highly influenced in short text scenario where the word co-occurrence patterns become very sparse in each document. To tackle this problem, we propose an improved CWTM model based on DMM, to leverage couple word co-occurrence to help distill better topics over short texts instead of the traditional word co-occurrence way.

3.1 Couple Word

The traditional topic model assumes that the topic is the distribution of a set of words, while finding the words related to each topic is achieved by acquiring the co-occurrence of words and words in each document. When a number of words always appear in the same document frequently, it can be said that these words belong to the same topic, which is also the method of getting the topic from the traditional topic model. However, considering the short text data sparseness and lack of semantic information, the traditional method of topic model is not reasonable for some theme polysemy, low-frequency word association and under the theme words problem of high degree of polymerization. Based on this consideration, the document after word segmentation is processed into unordered word pairs (e.g., topic/model/extraction can generate three couple words, namely topic/model, topic/extraction, model/extraction) and the word in each couple word is out of order. Supposing the two words in each couple word belong to the same topic which can not only solve the problem of polysemy, but also enrich the semantic information of short text. For example, for a short text "I bought a new-version of the Puma today". After removing the stop word, a new short text is obtained, that is to say "new-version/best-selling/puma". The probability that "new-edition" and "best-selling" fall on the same theme (clothes) is very high. However, Puma may belong to the topic of clothes and may also belong to topic of animal. If we generate a serious of unordered couple words, e.g., new-version/puma, best-selling/puma, through the interaction with the words in the couple word, "puma" is more likely to be split under the theme of clothing in this text. Thus, the extracted theme features can express this short text more than to express the topic of clothing rather than animal theme.

3.2 Our Approach

Topic extraction method of micro-blog short text proposed in this paper expands the semantic information of short text by extracting couple words, so as to alleviate the data sparseness problem of short text and improve the effect of topic features mining. Dirichlet Mixture Model [10] is a probabilistic topic model for restricted the document-topic distribution. That is, a document is generated from a single topic and all words within a document are generated by using the same topic distribution. We assume that the document is a mixture of topics. Each micro-blog short text consists of multiple couple word which is independent of each other. What's more, the word pairs are unordered and belong to the same topic. More formally, with the Dirichlet priors of α and β, the generative process of CWTM under the whole corpus is described as follows:

```
For all documents d ∈ {1,...,D} do
        sample mixture proportion  θ ~ Dirichlet(α)
        sample terms number  Nd ~ Poiss(ε)
        for all terms  n ∈ {1,...,Nd}, i  and  i ∈ {1,...,V}
        in document  d  do
        sample a topic  zd,n ~ Multinomail(θ)
        sample two words :  wi, wj ~ Multinomail(φzd,n)
For all topic  k ∈ [1,K]  do
        sample mixture distribution  φk ~ Dirichlet(β)
```

According to the above description, the model is converted to a probabilistic graph, as shown in Fig. 1.

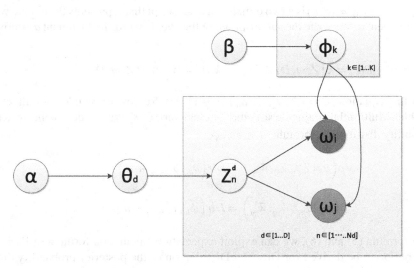

Fig. 1. The Bayesian network diagram of the method proposed in this paper. K is the number of topics, M is the total number of documents, and N_m is the number of couple words in document d. More formally, with the Dirichlet priors of α and β, $z_{d,n}$ is the topic of couple word n in document d, and w_i, w_j are the two words of the couple word. What's more, the implied variable θ_d represents the topic distribution under the document d which is a K dimensional vector, and the implied variable ϕ_k represents the word distribution under the topic k that is a V dimensional vector and V is the total number of words in the corpus.

According to the above description, given a collection of documents, w_i, w_j can be observed variables, α and β is given based on prior experience parameters, other variables of θ_d, ϕ_k, $z_{d,n}$ are unknown hidden variables, also according to the needs of the observed variables to learning evaluation. A joint probability density of all variables can be obtained from the Bayesian network graph in Fig. 1, as shown below:

$$p(\theta, \phi, z, b) = p(\theta/\alpha) \prod_{d=1}^{N_d} [p(\theta/\beta)p(z/\theta)p(\omega_i/z, \phi)p(\omega_j/z, \phi)] \tag{1}$$

where b is the number of couple words in document d, and the joint probability of b is:

$$p(b/\theta, \phi) = \sum_{k=1}^{K} [p(z = k/\theta)p(\omega_i/\phi)p(\omega_j/\phi)] \tag{2}$$

Thus the likelihood of the whole corpus is:

$$p(B/\theta, \phi) = \prod_{1}^{D} \prod_{1}^{N_d} p(b/\theta, \phi) \tag{3}$$

The definition of z_q represents the topic corresponding to the couple word q in the corpus z, where $q = (d, n)$ is a two dimensional subscript that represents the couple word n of document d. We write the subscript $\neg q$ for the couple words in document q removed, so:

$$P(z_q = k \mid Z\neg q, D) \propto P(z_q = k, d_q = (w_i, w_j) \mid Z\neg q, D) \tag{4}$$

In this equation, $D = (d_1, d_2, \ldots .d_m), m \in [1, M]$. Removing $\neg q$ does not affect the Dirichlet-Multiomial conjugate structure of other variables. We can derive the posterior probability distribution formula of θ_d and ϕ_k:

$$p\left(\vec{\theta}_d \mid \vec{z}_{\neg i}, \vec{w}_{\neg j}, \vec{w}_{\neg j} \right) = Dir\left(\vec{\theta}_d \mid \vec{\theta}_{d, \neg i, \neg j} + \vec{\alpha} \right) \tag{5}$$

$$p\left(\vec{\phi}_k \mid \vec{z}_{\neg q}, \vec{w}_{\neg j}, \vec{w}_{\neg j} \right) = Dir\left(\vec{\phi}_k \mid \vec{\theta}_{k, \neg i, \neg j} + \vec{\beta} \right) \tag{6}$$

In formula (5) and (6), we can exploit expectation calculation formula of Dirichlet Distribution to be derived for Dirichlet Distribution of the posterior probability distribution of parameters in the Bayesian framework as shown in Eqs. (7) and (8):

$$\theta_{dk} = \frac{n_{d,\neg q}^{(k)} + \alpha_k}{\sum_{k=1}^{K} \left(n_{d,\neg q}^{(w_i, w_j)} + \alpha_k \right)} \tag{7}$$

$$\phi_{kw} = \frac{n_{k,\neg q}^{w_i} + n_{k,\neg q}^{w_j} + \beta_k}{\sum_{1}^{V} \left(n_{k,\neg q}^{w_i} + n_{k,\neg q}^{w_j} + \beta_k \right)} \tag{8}$$

To this end, we propose a nonparametric probabilistic sampling strategy substitute formula (7) and formula (8) into formula (4) as follows:

$$P\left(z_q = k \mid \vec{Z}_{\neg q}, \vec{D} \right) \propto \frac{n_{d,\neg q}^{(k)} + \alpha_k}{\sum_{k=1}^{K} \left(n_{d,\neg q}^{(w_i,w_j)} + \alpha_k \right)} \cdot \frac{n_{k,\neg q}^{(w_i,w_j)} + \beta_k}{\sum_{1}^{N_d} \left(n_{k,\neg q}^{(w_i,w_j)} + \beta_k \right)} \tag{9}$$

This iterative process continues until the predefined number of iterations is reached.

4   Experiments and Results

In this section, we conduct extensive experiments to demonstrate the effectiveness of our proposed approach on real-world short text collections against the state-of-the-art alternative. The datasets used in the experiment are BaiduQA and Web Snipet of Sogou Labs. We performed the following preprocessing on the datasets: (1) remove the words with fewer than 3 characters; (2) convert letters to lowercase; (3) remove words with document frequency less than three in the dataset; (4) remove all non-alphabetic characters and stop words. After the above pretreatment, 89000 texts with moderate length were selected at random with an average length of 5 to 7 words. Among these, BaiduQA has 37,000 short texts with 30 categories while Web Snipet of Sogou Labs has 52,000 with 10 categories.

In addition, the word segmentation tool is the lCTCLAS of the Chinese Academy of Sciences, the classification tool uses the National Taiwan University LIBSVM, the code of DMM model experimental is from the literature, and LDA model experimental code is from the JAVA version of the Internet. All the experiments were run on a Linux server with Intel Xeon 2.33 GHz CPU and 16G memory. For all the methods in comparison, we set the hyperparameters $\alpha = K/50$ and $\beta = 0.01$ unless explicitly specified elsewhere. The number of iterations for Gibbs sampling is 1000 times.

Because the method proposed in this paper is based on DMM, the comparison method of this experiment is also DMM. In addition, the traditional topic model LDA is also chosen as a contrast method. The results of the experimental results are based on the effect of text classification in three evaluation indexes (Precision, Recall, and F-measure). The results of 10 times of running are shown in Table 1 according to different topic number of data set.

Table 1. Methods comparison results on the data set with the different topic number.

	K = 20				K = 40		
	Precision	Recall	F-measure		Precision	Recall	F-measure
CWTM							
BaiduQA	39.24%	47.88%	42.25%	BaiduQA	41.55%	50.78%	47.21%
Sougou	35.36%	45.77%	38.65%	Sougou	47.63%	31.85%	39.89%
DMM							
BaiduQA	35.54%	38.23%	32.38%	BaiduQA	36.52%	38.96%	34.80%
Sougou	40.87%	37.45%	35.52%	Sougou	35.21%	32.44%	30.12%
LDA							
BaiduQA	41.27%	37.50%	34.20%	BaiduQAA	39.68%	38.11%	36.14%
Sougou	36.88%	22.60%	29.80%	Sougou	33.19%	39.45%	32.40%

Figures 2 and 3 show the experimental result comparison of three methods on two data sets with different topic numbers, respectively. It can be intuitively found from the experimental result of the extraction algorithm on the BaiduQA data set in Fig. 2 that CWTM has the best performance, followed by LDA and DMM. The experimental result of the extraction algorithm on the Sougou data set in Fig. 3 also reflects that the test result of CWTM is the best. Except the poor performance on the data set with the topic number of less than 20, LDA has better performance on other two data sets than DMM.

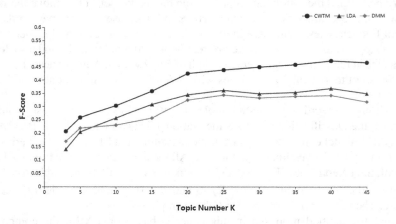

Fig. 2. The experimental result comparison of three methods on BaiduQA data set with different topic numbers.

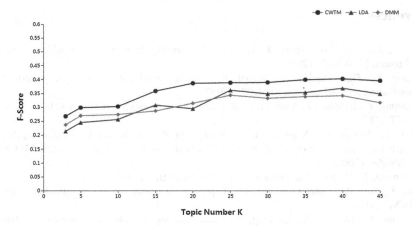

Fig. 3. The experimental result comparison of three methods on Sougou data set with different topic numbers.

The conclusion can be drawn from the analysis of Figs. 2 and 3: the topic extraction algorithm that leverages couple word co-occurrence to help distill better topics over short texts instead of the traditional word co-occurrence way based on Dirichlet Multinomial Mixture can obtain stable and good effect in performance.

5   Conclusions and Future Work

In this paper, we propose a novel topic model for short texts, named CWTM based on DMM, to leverage couple word co-occurrence to help distill better topics over short texts instead of the traditional word co-occurrence way. Although we do not introduce new topic models to address the issues of short text modeling especially in microblogging environments in this paper, our work sheds some light on how research on topic models can be conducted for short text scenarios. We conduct extensive experiments to demonstrate the effectiveness of our proposed approach on real-world short text collections against the state-of-the-art alternative. CWTM has good topic identification performance.

To the best of our knowledge, we leverage couple word co-occurrence to help distill better topics over short texts instead of the traditional word co-occurrence way for general short texts. However, there is still room to improve our work in the future. For example, we intend to validate the effectiveness of using word embedding techniques to help distill better topics over short texts. Topic extraction of complex structure has been always a difficulty in short text. The model mainly focus on solving the short problem. But the micro-blog data also have other problems, such as noises, and the diversity of language usage. Thus, the new algorithm proposed in the paper will be used in the research of this issue in the future.

References

1. Weng, J., Lim, E.-P., Jiang, J., He, Q.: Twitterrank: finding topic-sensitive influential twitterers. In: WSDM (2010)
2. Wang, X., Zhai, C., Hu, X., Sproat, R.: Mining correlated bursty topic patterns from coordinated text streams. In: SIGKDD (2007)
3. Xiaohui, Y., Jiafeng, G., Yanyan, L.: A biterm topic model for short texts. In: WWW, pp. 13–17 (2003)
4. Blei, D., McAuliffe, J.: Supervised topic models. In: Platt, J., Koller, D., Singer, Y., Roweis, S. (eds.) Advances in Neural Information Processing Systems 20, pp. 121–128. MIT Press, Cambridge (2008)
5. Hofmann, T.: Probabilistic latent semantic indexing. In: SIGIR (1999)
6. Ma, Z., Sun, A., Yuan, Q., Cong, G.: Topic-driven reader comments summarization. In: CIKM (2012)
7. Ramage, D., Dumais, S., Liebling, D.: Characterizing microblogs with topic models. In: International AAAI Conference on Weblogs and Social Media, vol. 5, pp. 130–137 (2010)
8. Rosen-Zvi, M., Griffiths, T., Steyvers, M., Smyth, P.: The author-topic model for authors and documents. In: UAI (2004)
9. Chen, J., Nairn, R., Nelson, L., Bernstein, M., Chi, E.: Short and tweet: experiments on recommending content from information streams. In: Proceedings of the 28th International Conference on Human Factors in Computing Systems, pp. 1185–1194. ACM (2010)
10. Wang, Y., Agichtein, E., Benzi, M.: TM-LDA: efficient online modeling of latent topic transitions in social media. In: Proceedings of the 18th ACM SIGKDD, New York, pp. 123–131. ACM (2012)
11. Sriram, B., Fuhry, D., Demir, E., Ferhatosmanoglu, H., Demirbas, M.: Short text classification in twitter to improve information filtering. In: SIGIR (2010)
12. Deerwester, S., Dumais, S., Furnas, G., Landauer, T., Harshman, R.: Indexing by latent semantic analysis. J. Am. Soc. Inf. Sci. **41**(6), 391–407 (1990)
13. Blei, D.M., Ng, A.Y., Jordan, M.I.: Latent dirichlet allocation. J. Mach. Learn. Res. **3**, 993–1022 (2003)
14. Teh, Y.W., Jordan, M.I., Beal, M.J., Blei, D.M.: Hierarchical dirichlet processes. J. Am. Stat. Assoc. **101** (2004)
15. Ramage, D., Dumais, S.T., Liebling, D.J.: Characterizing microblogs with topic models. In: ICWSM (2010)
16. Quan, X., Kit, C., Ge, Y., Pan, S.J.: Short and sparse text topic modeling via self-aggregation. In: AAAI (2015)
17. Lin, C.X., Zhao, B., Mei, Q., Han, J.: PET: a statistical model for popular events tracking in social communities. In: Proceedings of the 16th ACM SIGKDD, pp. 929–938. ACM (2010)
18. Weng, J., Lim, E., Jiang, J., He, Q.: Twitterrank: finding topic-sensitive influential twitterers. In: Proceedings of the Third ACM International Conference on Web Search and Data Mining, pp. 261–270. ACM (2010)
19. Zhai, K., Boyd-Graber, J.L.: Online latent dirichlet allocation with infinite vocabulary. In: ICML, vol. 28, no. 1, pp. 561–569 (2013). JMLR Proceedings. JMLR.org
20. Zhao, W., Jiang, J., Weng, J., He, J., Lim, E., Yan, H., Li, X.: Comparing twitter and traditional media using topic models. In: Advances in Information Retrieval, pp. 338–349 (2011)
21. Phelan, O., McCarthy, K., Smyth, B.: Using twitter to recommend real-time topical news. In: Proceedings of the Third ACM Conference on Recommender Systems, New York, pp. 385–388. ACM (2009)

22. Hong, L., Davison, B.: Empirical study of topic modeling in twitter. In: Proceedings of the First Workshop on Social Media Analytics, pp. 80–88. ACM (2010)
23. Röder, M., Both, A., Hinneburg, A.: Exploring the space of topic coherence measures. In: Proceedings of the Eighth ACM International Conference on Web Search and Data Mining, pp. 399–408. ACM (2015)

Micro-blog User Community Detection by Focusing on Micro-blog Content and Community Structure

Jia Liu[1(✉)], Ya-jun Du[1], and Ji-zhong Ren[2]

[1] School of Computer and Software Engineering,
Xihua University, Chengdu 610039, China
`xiaoke92@foxmail.com`
[2] School of Political and Public Administration,
Electronic Science and Technology University, Chengdu 611731, China

Abstract. Micro-blog community detection is one of the hot problems of Micro-blog platform. There are many existing community detection methods that are dedicated to detect community by only considering the topological structure. To detect Micro-blog community better, we considering the Micro-blog content as well as the topological structure. In Micro-blog community, the essence of a concept is semantic objects in the real world. The concept is composed of the object's attribute set, and the attribute set is a set of nouns that essentially can represent the object. In this article, we let user be object and calculate the interest similarity by the object's attribute set. First, we establish a micro-blog social network by analyzing the object's attribute set. Second, we find the clustering directions for each object by the Random Walk method. Then, we detect micro-blog user community following the clustering directions. Finally, experiments performed to verify the efficiency of our method from the two aspects of community structure and interest cohesion.

Keywords: Community detection · Micro-blog content · Random walk

1 Introduction

In recent years, the researches that focus on the Micro-blog platform mainly include: Micro-blog user behavior analysis [1], Micro-blog language analysis and identification, standard word processing [2], information dissemination and public opinion model [3], and Micro-blog community division and detection. Micro-blog community detection is the main research work. The objective of a community detection is to identify groups of nodes that have common interests, habits and hobbies in a specific network. Many algorithms have been proposed to find communities, such as the local network community detection method [4], the cavity method [5], and the parallel community detection algorithm [6]. The vast majority of those methods concentrate on community structure to detect a community.

© Springer Nature Singapore Pte Ltd. 2017
J. Li et al. (Eds.): CCKS 2017, CCIS 784, pp. 92–103, 2017.
https://doi.org/10.1007/978-981-10-7359-5_10

A micro-blog is the epitome of social reality because it provides people a huge amount of valuable data concerning micro-blog contents. To detect a community in the micro-blog network, we let user be object and calculate the interest similarity by the object's attribute set. Different than the existing methods, we combine community structure and micro-blog contents to detect a community. Detecting a community in a social network raises new challenges to traditional community detection algorithms and is the basis for studying the knowledge map that based on Micro-blog community. To analyze a social network community, higher rationality and adaptability force us to mine user preferences. The entire process of our method is shown in Fig. 1. Our main contributions in this paper can be summarized as follows:

(1) The micro-blog social network is described as a tuple $(\mathrm{MSN} = \langle O, R, C \rangle)$. Then, we let user be object and calculate the interest similarity by the object's attribute set.
(2) Detect the clustering directions based on the Random Walk method.
(3) We detect communities in the micro-blog network by following the clustering directions.
(4) We analyze the performance of our method via experiments from two aspects: community interest cohesion and community structure.

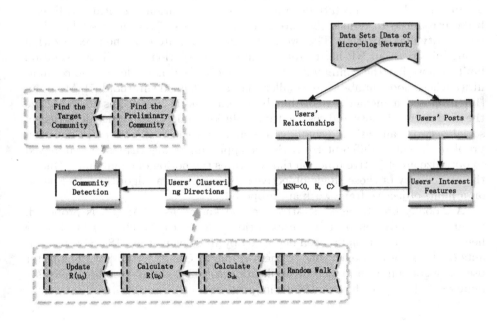

Fig. 1. The entire process of detecting micro-blog user community.

The structure of the paper is as follows: In the next section, we present a review of the literature. In Sect. 3, we provide a detailed discussion of building a micro-blog network graph. In Sect. 4, we will find clustering directions for each object. Community detection will be shown in Sect. 5. Finally, we present the results of our experiments in Sect. 6.

2 Related Work

According to the characteristics of network structure [7], community detection methods can be roughly categorized into positive network community detection and signed network community detection. Moreover, among efficient heuristics for community detection we can distinguish between those based on community agglomeration and those based on local node moves.

The Spectral Bisection algorithm [8] and the Kernighan-Lin algorithm [9] are two classical algorithms, which have been recommended for detecting positive network communities. Brandes et al. [10] showed the NP-complete problem of the Spectral Bisection algorithm. Additionally, the variations in the order of the partitions may significantly alter the results because of the sensitivity of the Kernighan-Lin algorithm to the initial partition. Several algorithms based on optimizing the modularity are developed to detect the community structures of complex networks, especially for weighted networks and directed networks [11] to address the issues of these two classical methods [12,13]. Furthermore, to obtain the suboptimal solutions, a few of the optimization algorithms [14] have been introduced to reduce the two major limitations of the methods based on modularity maximization. The two major limitations are that the maximization of modularity is an NP-hard problem and the resolution limit. The resolution limit means that the community detection methods cannot detect the communities whose node numbers are smaller than a predefined threshold [7]. Mu et al. [15] proposed a memetic algorithm based on genetic algorithms to maximize the modularity density and resolve the resolution limit. Community detection in social network analysis is usually considered as a single-objective optimization problem, in which different heuristics or approximate algorithms are employed to optimize an objective function that captures the notion of community. Due to the inadequacy of those single-objective solutions, several algorithms [16] based on a multi-objective framework are proposed.

A globally greedy agglomerative method known as CNM [17] is proposed. Community detection by label propagation belongs to the class of local move heuristics. It has originally been described by Raghavan et al. [18]. Several variants of the algorithm exist, one of them is due to Gilbert et al. [19]. The latter use the algorithm as a prototype application within a parallel toolbox that uses numerical algorithms for combinatorial problems.

3 Establish the Micro-blog Network

We describe a Micro-blog Social Network (MSN) by a tuple (MSN = $\langle O, R, C \rangle$), where O represents a set of users or objects, R is a set of objects' relationships and C denotes a set of groups of objects. Each group $c_i \in C(i = 1,2,3...)$ is a subset of O.

Based on the majority of traditional community detection methods that focus on the social network topology, we consider the content analysis of the MSN in our approach. In our method, $O = \{u_1, u_2, \cdots, u_i, \cdots\}$ represents a set of objects, $u_i = (\boldsymbol{p_i})$ is an object, and $\boldsymbol{p_i}$ is an interest feature vector. If object u_i follows object u_j, there is an edge $e_{ij} \in R$ between u_i and u_j. We define w_{ij} as the interest similarity between two objects, u_i and u_j. Simultaneously, we allow w_{ij} be the weight of the edge e_{ij}. To compute the w_{ij}, we improve the Tanimoto coefficient. The improved Tanimoto coefficient formula is as follows:

$$w_{ij} = \frac{\boldsymbol{p_i}' \bullet \boldsymbol{p_j}'}{|\boldsymbol{p_i}'|^2 + |\boldsymbol{p_j}'|^2 - \boldsymbol{p_i}' \bullet \boldsymbol{p_j}'}. \tag{1}$$

Where $\boldsymbol{p_i}'$ is a dimension-oriented Boolean expression of $\boldsymbol{p_i}$, which can be obtained by incorporating features and changing the type of features to Boolean. Formula (2) shows each element p_{ik}' of $\boldsymbol{p_i}'$. In formula (2), $\boldsymbol{p_i} \bar{U} \boldsymbol{p_j}$ represents a vector that contains all features belonging to the vector $\boldsymbol{p_i}$ or $\boldsymbol{p_j}$. In the $\boldsymbol{p_i} \bar{U} \boldsymbol{p_j}$, elements are not equal to each other. The formula for each element p_{ik}' of $\boldsymbol{p_i}'$ is as follows:

$$p_{ik}' = \begin{cases} 1, & \text{if } r_k \in \boldsymbol{p_i}, \\ 0, & \text{others.} \end{cases} \quad r_k \in \boldsymbol{p_i} \bar{U} \boldsymbol{p_j} \text{ and } k = 1 \cdots Count(\boldsymbol{p_i} \bar{U} \boldsymbol{p_j}). \tag{2}$$

where r_k is an element of the $\boldsymbol{p_i} \bar{U} \boldsymbol{p_j}$. The function Count(Vector e) is used to calculate the count of the element in vector e. Then, $\boldsymbol{p_i}'$ can be represented as $\boldsymbol{p_i}' = (p_{i1}', p_{i2}', \cdots, p_{ik}')$.

4   Detect the Objects' Clustering Directions

We adopt characteristics of a social network to determine the clustering directions for each object. We use game theory to quantify the three characteristics of social networks. We take four steps to detect the clustering directions.

To calculate the decision for random walking steps, we utilize a possibility function proposed by Xin et al. [20]. They designed the random walk process as follows: $a \in (0,1)$, f(x), where x is the current step of the walker and f(x) is the possibility function for deciding whether to continue walking or not. At the current walking step, if $a < f(x)$, the walker performs the next step, otherwise, terminates walking. The random walk possibility function is the following:

$$f(x) = \frac{1}{\exp((x - \frac{1}{\eta d^\omega + 1})^\delta - \beta) + 1} \tag{3}$$

Where δ controls the distribution of the walking steps, and β controls the furthest walking steps. Because of the 'Six Degrees of Separation', the diameter of the community is less than 6. Thus, the value of β should be 6. The formula $x - \frac{1}{\eta d^\omega + 1}$ is a topology gain function based on the logistic function. In formula (3), d is the degree of the object. ω and η are the adjustable parameters of the logistic function.

In the second step, we adopt the game theory to quantify the importance of objects to the original object. Additionally, we can obtain more useful objects for the origin object by this method. The random walking earning and penalty function is as follows:

$$S_{u_k} = \varphi(\sum w_{ij} - \beta' * S_{loss}) + (1 - \varphi)(1 - \frac{Path_{min}(u_0, u_{ter})}{\beta}) \qquad (4)$$

$$S_{loss} = 1 - \frac{\sum\sum w_{kl}}{2n} \quad u_k, u_l \in O \bigwedge k \neq l \qquad (5)$$

Where S_{u_k} is the actual score of u_k that is one of the neighbors of the origin object. β' represents a practical walking step. The u_0 and u_{ter} represent the origin object and the terminal object for the walker, respectively. To prevent the walker from having repellency back to the origin object u_0, we design the formula, $1 - \frac{Path_{min}(u_0, u_{ter})}{\beta}$, to solve the problem. It means that if the shortest distance between the origin object and the terminal object is large, the walker has a low probability of returning to the origin object. That is the terminal object has little significance to the origin object. The function $Path_{min}(u_0, u_{ter})$ is designed to calculate the shortest distance between u_0 and u_{ter}. The parameter φ measures the importance of the similarity between objects and the walking trend of the walker. The formula (5) is a punishment value of each step.

In the third step, we utilize the results of N times random walking to find the clustering directions for each object. First, we determine one of the clustering directions u_k for the origin object u_0 by the following formula:

$$R(u_0) = \{u_k | u_k = \arg\max(\overline{S_{u_k}}) \bigwedge u_k \in Ne(u_0)\} \qquad (6)$$

Where $R(u_0)$ represents the set of the clustering directions for object u_0, and $\overline{S_{u_k}}$ represents the average score of n ($n \leq N$) times random walking for the walker through u_k. $Ne(u_0)$ denotes the set of neighbors of u_0. Formula (6) means that the object u_k, which has a maximal score, is one of the clustering directions for object u_0. Specially, objects that have a maximal average score of more than one are all the clustering directions of the object u_0.

In the fourth step, we will find additional clustering directions for object u_0. If object $u_{k'}$ has a score which is close to the maximal average score, we would determine that $u_{k'}$ is a clustering direction of object u_0. The clustering directions of object u_0 can be updated by the following formula:

$$R(u_0) = \{u_k, u_{k'}, u_{k''}, \cdots | u_k = \arg\max \overline{S_{u_k}}, (u_k - u_{k'}) < \varepsilon, (u_k - u_{k''}) < \varepsilon, \cdots\} \qquad (7)$$

Where the parameter ε is a small value. This shows that if the objects have a score close to object u_k, these objects would be the clustering directions of the object u_0.

5   Detect Community

After detecting the clustering directions for each object, we will detect community by two steps.

In the first step of detecting community, we will find some small communities by analyzing the similarity and clustering directions of objects. First, we define two sets (m_Set and s_Set) to show the clustering directions of two objects. If two objects have a mutual clustering direction, they belong to the m_Set. However, the two objects will belong to the s_Set if they have a single-track clustering direction. Then, we sort the two sets according to the similarity of the two objects. Finally, if two objects in the m_Set or s_Set do not belong to any community, we will create a new community for them. In this process, we first consider the clustering directions in m_Set, and those in s_Set are reviewed later. Because the two objects, who have mutual clustering directions and a higher similarity, are more interested in a common community.

In the second step of detecting community, the target communities will be detected by considering the topological structure. First, if an element of the m_Set and s_Set has been formed into a small community, we delete it from the corresponding set. Each element in the two updated sets must contain an object u_i who already belongs to a community and another object u_j who is not member of any community. Then we will detect community for object u_j by analyzing the change of the topological structure. The modularity [21] has the unique privilege of also being a global criterion to assess the compactness of the community structure. The formula of modularity is shown in formula (8). Arab et al. [22] studied the modularity [21] in more depth. They [22] analyzed the change of the modularity when two communities unite to one. Therefore, we use the change of modularity as an evaluation criteria for the change of the topological structure when two communities unite into one. The formula for the change of the modularity is as follows:

$$Q = \sum_i (e_{ii} - a_i{}^2) \tag{8}$$

$$\Delta Q = \frac{E_{ij}}{m} - 2a_i a_j \tag{9}$$

Where Q is the modularity of a network, i is the number of a community, e_{ii} is the fraction of edges in community i, a_i is the fraction of edges that connect two vertices in community c_i, and ΔQ represents the change in the modularity value when two communities unite to a new community. The number of edges between the two communities, c_i and c_j, are denoted as E_{ij}. The edges of the entire social network are denoted as m. Assuming d_v is the degree of node v, a_i can be presented as $a_i = \frac{\sum_{v \in c_i} d_v}{\sum_{v \in MSN} d_v}$. Based on the m_Set and the s_Set, we will merge communities if $\Delta Q > \kappa (\kappa > 0)$. The parameter κ is a threshold for ΔQ. We merge communities following the decrease in ΔQ until the merging condition cannot be met.

6   Experiments

Two datasets which are labeled *Dataset1* and *Dataset2* are subject to experimentation in this study. We obtain the *Dataset1* from Sina micro-blog open platform and the *Dataset2* from the Micro-blog User Portrait Contest in 2016. Figures 2 and 3 show the network structure distribution of two datasets. The detailed information of the two datasets is shown in Table 1.

Our method is implemented in Python. We conduct our experiments on two datasets and contrast the results with other proposed algorithms, such as the CNM approach [17], Infomap [23], COPRA [24] and NRW [20]. The CNM approach [17] is a popular community detection approach and is currently widely used. Hence, we utilize CNM as one of the contrast approaches. The Infomap [23] and NRW [20] methods are both random-walks methods based on the community structure. Therefore, we chose them be our additional contrast approaches. Furthermore, our method not only considers community structure, but detects communities by analyzing micro-blog content. Therefore, we will analyze the performance of our method from two aspects: the community structure and community interest cohesion.

Table 1. The two datasets for the experiments.

Description	*Dataset1* (Size)	*Dataset2* (Size)
Followers and followees	311361	321712
User posts	43721	49921
Micro-blog users	781	980

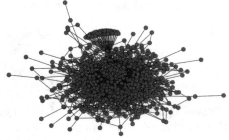

Fig. 2. The network structure distribution of *Dataset1*.

Fig. 3. The network structure distribution of *Dataset2*.

6.1   Evaluate by the Interest Cohesion

The objects in a common community have a highly similar interest in general. Thus, we contrast our method with other methods using an evaluating indicator, e.g., interest cohesion. The interest cohesion $coi(c_i)$ of a community c_i can be calculated by formula (10).

$$coi(c_i) = \frac{\sum_{u_j \in c_i, u_k \in c_i} w_{jk}}{\sum_{u_j \in MSN, u_k \in MSN} w_{jk}} \quad (u_j \in Ne(u_k)) \tag{10}$$

Table 2 shows the comparison of the interest cohesion for the methods mentioned above. In Table 2, $Num(Cs)$ represents the number of communities. Our method produces a $Num(Cs)$ value close to the NRW and Infomap methods. Because these three methods are all based on the random walking method. The $Num(Os')$ denotes the number of objects in a community that has a maximal number of objects, and $Os' \in \max Num(Cs)$. From the values of $Num(Os')$, we can see that the CNM method has maximal number of objects in a community than other methods. The $\sqrt{s(coi(c_i))}$ represents the statistical dispersion of the interest cohesion. The maximal values of $Num(Os')$ and $\sqrt{s(coi(c_i))}$ show that the communities detected by the CNM method have an unbalanced distribution, i.e., there is a community c_i far greater than another community c_j from the number of objects. In contrast, our method has a lower value in $\sqrt{s(coi(c_i))}$. Thus, our communities have a balanced distribution. The $\overline{coi(c_i)}$ is an average interest cohesion and measures the goodness of communities from their interest similarity. Our method has the highest similarity value, because our method considers interest similarity when it detects a community. The interest cohesion of communities in $Dataset2$ is higher than in $Dataset1$, because $Dataset2$ is pre-processed artificially.

Table 2. The comparison of the community interest cohesion for the methods mentioned above. In the $Num(Os')$, $Os' \in \max Num(Cs)$, the $Num(Os')$ denotes the number of objects in a community that has a maximal number of objects.

	Dataset1				
	CNM [17]	COPRA [24]	Infomap [23]	NRW [20]	Our method
$Num(Cs)$	37	89	51	58	64
$Num(Os')$	213	80	97	105	81
$\overline{coi(c_i)}$ ($\times 10^{-2}$)	0.5557	0.1217	0.6863	0.5388	0.7227
$\sqrt{s(coi(c_i))}$ ($\times 10^{-2}$)	001.1024	0.0351	0.5545	0.5387	0.0295
	Dataset2				
	CNM [17]	COPRA [24]	Infomap [23]	NRW [20]	Our method
$Num(Cs)$	28	65	49	39	46
$Num(Os')$	271	117	178	137	142
$\overline{coi(c_i)}$ ($\times 10^{-2}$)	0.9212	0.5667	0.8333	0.8921	1.1721
$\sqrt{s(coi(c_i))}$ ($\times 10^{-2}$)	0.7291	0.0313	0.4923	0.5127	0.0301

We sort the communities following the number of objects to better compare the methods mentioned above. Figure 4(a) and (b) show the distribution the objects' number in the top-ten ranked communities. In the CNM method, the number of objects in communities has an unbalanced distribution. The number of objects in the first-ranked community is roughly twice as high as that in the second-ranked community. The other four methods have a similar distribution of the number of objects. Figure 5(a) and (b) show the distribution of the interest cohesion in the top-ten ranked communities. We can find that the community interest cohesion in our method is greater than the other methods. The communities with a large number of objects have high interest cohesion. The first-ranked community in our method has a smaller number of users than that in the CNM method. However, the first-ranked community in our method has higher interest cohesion than that in the CNM method. This means that the objects in communities in our method have higher interest cohesion.

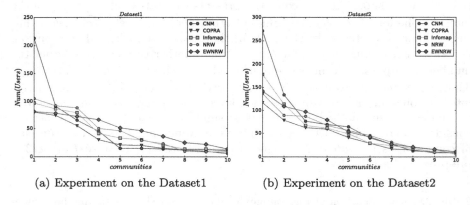

(a) Experiment on the Dataset1   (b) Experiment on the Dataset2

Fig. 4. Distribution of the number of objects in the top-ten ranked communities.

6.2   Evaluate by the Community Structure

Newman et al. [12] proposed a quality measure called modularity Q to quantify the community quality of a network or graph structure. The higher the value of modularity Q, the stronger the community structure is. Therefore, we adopt the modularity Q to analyze the performance of our method. The formula of modularity is shown in formula (8).

Table 3 shows the comparison of the community modularity for the methods mentioned above. In Table 3, the CNM method has a maximal modularity for the entire network and has a maximal single community modularity. This demonstrates that the objects are densely connected in an intra-community from the community structure. Moreover, our method has an approximate modularity with other methods, and thus, our method has good performance in the community structure. Figure 6(a) and (b) show the distribution of the modularity in the top-ten ranked communities. We find that the CNM method has the largest modularity and our method has high modularity. Therefore, our method is effective in detecting communities from the community structure.

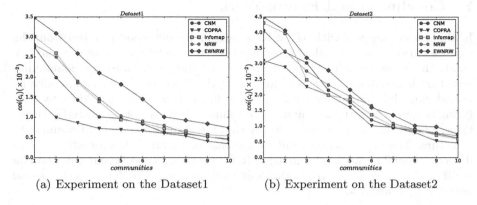

(a) Experiment on the Dataset1   (b) Experiment on the Dataset2

Fig. 5. Distribution of the community interest cohesion in the top-ten ranked communities.

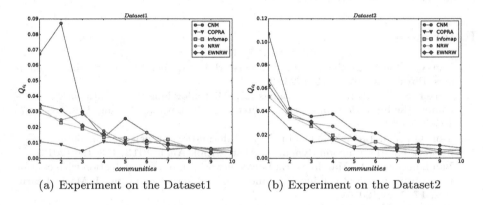

(a) Experiment on the Dataset1   (b) Experiment on the Dataset2

Fig. 6. Distribution of the modularity in the top-ten ranked communities.

Table 3. The comparison of community modularity for the methods mentioned above.

	Dataset1				
	CNM [17]	COPRA [24]	Infomap [23]	NRW [20]	Our method
$\max Q_{c_i}$	0.0872	0.0110	0.0323	0.0294	0.0345
$\sum_i Q_{c_i}$	0.3680	0.1667	0.1912	0.2459	0.2572
	Dataset2				
	CNM [17]	COPRA [24]	Infomap [23]	NRW [20]	Our method
$\max Q_{c_i}$	0.1072	0.0467	0.0528	0.0671	0.0625
$\sum_i Q_{c_i}$	0.5124	0.3023	0.3741	0.3908	0.4099

7   Conclusion and Future Work

In this paper, we deal with Micro-blog community detection problem from the topological structure and the Micro-blog content. We let user be object and calculate the interest similarity by the object's attribute set. Moreover, community structure is an important aspect in a network. Thus, we detect community by considering the community structure as well. First, we build the MSN by analyzing the objects' attributes and relationships. Then, we detect clustering directions. Finally, we detect community by the clustering directions and community structure. The experimental results show that our method has an advantage in detecting social network communities. In future work, we will use our method on additional data sets and verify that our method is superior for detecting social network communities.

Acknowledgments. This work is supported by the National Nature Science Foundation (Grant No. 61271413, 61472329 and 61532009) and the Innovation Fund of Postgraduate, Xihua University.

References

1. Yan, Q., Wu, L.R., Zheng, L.: Social network based microblog user behavior analysis. Physica A **392**, 1712–1723 (2013)
2. Carter, S., Weerkamp, W., Tsagkias, M.: Microblog language identification: overcoming the limitations of short, unedited and idiomatic text. Lang. Res. Eval. **47**, 195–215 (2013)
3. Yan, Q., Wu, L.R., Liu, C., Li, X.Y.: Information propagation in online social network based on human dynamics. Abstr. Appl. Anal. **2013**, 173–186 (2013)
4. Van Laarhoven, T., Marchiori, E.: Local network community detection with continuous optimization of conductance and weighted kernel K-means. J. Mach. Learn. Res. **17**, 1–28 (2016)
5. Zhang, P., Moore, C., Newman, M.E.J.: Community detection in networks with unequal groups. Phys. Rev. E **93**, 012303 (2016)
6. Palsetia, D., Hendrix, W., Lee, S., Agrawal, A., Liao, W., Choudhary, A.: Parallel community detection algorithm using a data partitioning strategy with pairwise subdomain duplication. In: Kunkel, J.M., Balaji, P., Dongarra, J. (eds.) ISC High Performance 2016. LNCS, vol. 9697, pp. 98–115. Springer, Cham (2016). https://doi.org/10.1007/978-3-319-41321-1_6
7. Guo, W.F., Zhang, S.W.: A general method of community detection by identifying community centers with affinity propagation. Physica A **447**, 508–519 (2016)
8. Fiedler, M.: Algebraic connectivity of graphs. Czech. Math. J. **23**, 298–305 (1973)
9. Kernighan, B.W., Lin, S.: An efficient heuristic procedure for partitioning graphs. Bell Syst. Tech. J. **49**, 291–307 (1970)
10. Brandes, U., Delling, D., Gaertler, M., Görke, R., Hoefer, M., Nikoloski, Z., Wagner, D.: On finding graph clusterings with maximum modularity. In: Brandstädt, A., Kratsch, D., Müller, H. (eds.) WG 2007. LNCS, vol. 4769, pp. 121–132. Springer, Heidelberg (2007). https://doi.org/10.1007/978-3-540-74839-7_12

11. Li, Y.D., Liu, J., Liu, C.L.: A comparative analysis of evolutionary and memetic algorithms for community detection from signed social networks. Soft. Comput. **18**, 329–348 (2014)
12. Newman, M.E.J., Girvan, M.: Finding and evaluating community structure in networks. Phys. Rev. E **69**, 026113 (2004)
13. Wu, J.S., Wang, F., Xiang, P.: Automatic network clustering via density-constrained optimization with grouping operator. Appl. Soft Comput. **38**, 606–616 (2016)
14. Atay, Y., Koc, I.: Community detection from biological and social networks: a comparative analysis of metaheuristic algorithms. Appl. Soft Comput. **50**, 194–211 (2016)
15. Mu, Ca.H., Xie, J., Liu, Y.: Memetic algorithm with simulated annealing strategy and tightness greedy optimization for community detection in networks. Appl. Soft Comput. **34**, 485–501 (2015)
16. Zou, F., Chen, D.B.: Community detection in complex networks: multi-objective discrete backtracking search optimization algorithm with decomposition. Appl. Soft Comput. **53**, 285–295 (2017)
17. Clauset, A., Newman, M.E.J., Moore, C.: Finding community structure in very large networks. Phys. Rev. E **70**, 066111/1–066111/6 (2004)
18. Raghavan, U.N., Albert, R., Kumara, S.: Near linear time algorithm to detect community structures in large-scale networks. Phys. Rev. E Stat. Nonl. Soft Matter Phys. **76**, 036106 (2007)
19. Gilbert, J.R., Reinhardt, S., Shah, V.B.: High-performance graph algorithms from parallel sparse matrices. In: Kågström, B., Elmroth, E., Dongarra, J., Waśniewski, J. (eds.) PARA 2006. LNCS, vol. 4699, pp. 260–269. Springer, Heidelberg (2007). https://doi.org/10.1007/978-3-540-75755-9_32
20. Xin, Y., Xie, Z.Q., Yang, J.: The adaptive dynamic community detection algorithm based on the non-homogeneous random walking. Physica A **450**, 241–252 (2016)
21. Newman, M.E.J.: Modularity and community structure in networks. Proc. Natl. Acad. Sci. **103**, 8577–8582 (2006)
22. Arab, M., Afsharchi, M.: Community detection in social networks using hybrid merging of sub-communities. J. Netw. Comput. Appl. **40**, 73–84 (2014)
23. Rosvall, M.: Maps of random walks on complex networks reveal community structure. Proc. Natl. Acad. Sci. **105**, 1118–1123 (2008)
24. Gregory, S.: Finding overlapping communities in networks by label propagation. New J. Phys. **12**, 2011–2024 (2010)

Embedding Syntactic Tree Structures into CNN Architecture for Relation Classification

Feiliang Ren[1,2(✉)], Rongsheng Zhao[1,2], Xiao Hu[1,2],
Yongcheng Li[1,2], Di Zhou[1,2], and Cunxiang Wang[1,2]

[1] School of Computer Science and Engineering, Northeastern University,
Shenyang, 110819, China
renfeiliang@cse.neu.edu.cn
[2] Key Laboratory of Medical Image Computing of Ministry of Education,
Northeastern University, Shenyang, 110819, China

Abstract. Relation classification is an important task in natural language processing (NLP) fields. State-of-the-art methods are mainly based on deep neural networks. This paper proposes a new convolutional neural network (CNN) architecture which combines the syntactic tree structure and other lexical level features together for relation classification. In our method, each word in the input sentence is first represented as a k-size word sequence which contains the context information of the considering word. Then each of such word sequence is parsed into a syntactic tree structure and this kind of tree structure is further mapped into a real-valued vector. Finally, concatenated with the attention features for the words among the marked entities, all of these features are fed into a CNN model for relation decision. We evaluate our method on the SemEval 2010 relation classification task and experimental results show that our method outperforms previous state-of-the-art methods under the condition of without using external linguistic resources like WordNet.

Keywords: Relation classification · Syntactic parsing tree
Convolutional neural networks

1 Introduction

The aim of relation classification is to select a proper semantic relation type from a predefined set of relations for two labelled entities in a given input sentence. For example, given a sentence "*The system as described above has its greatest application in an arrayed <e1> configuration </e1> of antenna <e2> elements </e2>*", a relation classification system aims to identify that there is a "*Component-Whole*" relation from entity e_2 to e_1. Relation classification is an important fundamental research in NLP fields. Accurate relation classification results would benefit lots of NLP tasks, such as sentence interpretations, question & answering, knowledge graph construction, ontology learning, and so on. Thus, lots of researchers have devoted to relation classification research in recent years.

© Springer Nature Singapore Pte Ltd. 2017
J. Li et al. (Eds.): CCKS 2017, CCIS 784, pp. 104–116, 2017.
https://doi.org/10.1007/978-981-10-7359-5_11

Early research mostly focused on features based methods. These methods firstly select some effective features of the given sentence from syntactic and semantic aspects, then the selected features are inputted into some classification models like support vector machines or maximum entropy. Recently, deep neural network (DNN) based methods have been used in relation classification task and have achieved state-of-the-art experimental results. The core of these methods is to embed lots of useful features (distance features, word string features, etc.) into vectors, and then input these vectors into some deep neural network architectures for final relation decisions. Usually, deep convolutional neural networks (CNN) and deep recurrent neural networks (RNN) are two of the most widely used DNN architectures for relation classification.

Previous research (Bach and Badaskar 2007) has shown that syntactic tree structures would be of great help for relation classification, thus more and more research attentions have been paid to integrating syntactic tree structures into DNN models. However, in these existing methods, syntactic tree structures are used in a very shallow manner: syntactic tree is taken as an intermediate technology whose aim is to select a short fragment from the original sentence, on which a CNN model or a RNN model can be performed. Such shallow manner does not make full use of the syntactic or semantic richness of phrases in a sentence. To address this problem, Socher et al. (2013a, b) introduced a Compositional Vector Grammar (CVG), which used a syntactically untied RNN model to learn a syntactic-semantic compositional vector representation for the category nodes in a syntactic tree. Inspired by this work, this paper proposes a new relation classification method that combines the syntactic tree structures into a CNN model. We first select a local context window for each word in the input sentence and the word sequence in this context window is then parsed into a syntactic tree. Using the method introduced by Socher et al. (2013a, b), each of such syntactic trees is encoded into a real-valued vector. Concatenated with word embedding features and position vector features, the final feature vector for each word would be formed. Finally, these features are fed into a CNN model to predict the relation for the two labelled entities in a sentence.

2   Related Work

Our work is inspired by two kinds of research: learning vector representations for syntactic tree structures and relation classification with DNN models.

- Learning Vector Representations for Syntactic Tree Structures

The nature of learning vector representation for syntactic tree structures is to capture the full syntactic and semantic richness of linguistic phrases. The original research effort for this is to improve discrete representations for parsing. Researchers have proposed several approaches for this aim. However, most previous work attempts to improve on this by lexicalizing phrases or splitting categories. For example, Petrov et al. (2007) used a learning algorithm to split and merge the syntactic categories. Socher et al. (2013a, b) pointed out that such efforts can only partly address the problem at the cost of huge feature spaces and sparseness. They extend these previous ideas from discrete representations to richer continuous ones. Their CVG can be seen as factoring discrete and continuous parsing in one model (Fig. 1 is a CVG example shown in their paper). Their research directly inspires our work. We hope the performance of relation classification can be further improved by integrating the vector representations for syntactic tree structures into a CNN model.

- DNN based Models for Relation Classification

Generally, there are three widely used DNN architectures in relation classification: CNN, RNN, and their combination.

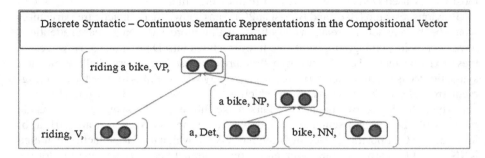

Fig. 1. Example of a CVG tree with (category, vector) representations at each node.

Zeng et al. (2014) proposed a CNN based approach for relation classification. In their method, sentence level features are learned through a CNN model that takes word embedding features and position embedding features as input. In parallel, lexical level features are extracted from some context windows that are around the labelled entities. Then sentence level and lexical level features are concatenated into a single vector. This vector is then fed into a *softmax* classifier for relation prediction. Dos Santos et al. (2015) also used the CNN model for relation classification. In their work, they proposed a ranking loss function with data cleaning. Wang et al. (2016) proposed a multi-level attention CNN model for relation classification. In their method, two levels of attention are used in order to better discern patterns in heterogeneous contexts. Nowadays, many works concentrate on extracting features from the *Shortest Dependent Path* (SDP). Xu et al. (2015a) learned robust relation representations from SDP through a CNN, and proposed a straightforward negative sampling strategy to improve the assignment of subjects and objects.

Socher et al. (2012) used RNN for relation classification. In their method, they build recursive sentence representations based on syntactic parsing. In contrast, Zhang and Wang (2015) investigated a temporal structured RNN with only words as input. They used a bi-directional model with a pooling layer on top. Xu et al. (2015b) picked up heterogeneous information along the left and right sub-path of the Shortest Dependent Path (SDP) respectively, leveraging RNN with LSTM. In their method, the SDP retains most relevant information to relation classification, while eliminating irrelevant words in the sentence. And the multichannel LSTM networks allow effective information integration from heterogeneous sources over the dependency paths. Meanwhile, a customized dropout strategy regularizes the neural network to alleviate over-fitting. There are some other research efforts. For example, Hashimoto et al. (2013) explicitly weighted phrases' importance in RNNs to improve performance. Ebrahimi and Dou (2015) rebuilt an RNN on the dependency path between two labelled entities.

Some researchers also combined CNN and RNN for relation classification. For example, Vu et al. (2016) investigated CNN and RNN as well as their combination for relation classification. They proposed extended middle context, a new context representation for CNN architecture. The extended middle context uses all parts of the sentence (the relation arguments, left/right and between of the relation arguments) and pays special attention to the middle part. They also presented a connectionist bi-directional RNN model and a ranking loss function is introduced for the RNN model. Finally, CNN and RNN were combined with a simple voting scheme. Cai et al. (2016) proposed a bidirectional neural network BRCNN, which consists of two RCNNs that can learn features along SDP inversely at the same time. Specifically, information of words and dependency relations are used with a two-channel recurrent neural network with LSTM units. The features of dependency units in SDP are extracted by a convolution layer. Liu et al. (2015) used a RNN to model the sub-trees, and a CNN to capture the most important features on the SDP.

3 Our Method

Figure 2 demonstrates the architecture of our method. The network takes sentences as input and extracts the syntactic tree features and other useful features. These features will be converted into vector representations and be fed into a CNN model for relation decision. There are five main components in our methods: feature extraction, convolution operation, max pooling operation, linear transformation, and output.

3.1 Feature Extraction

Given an input sentence $s = w_1w_2w_3...w_n$ with two marked entities e_1 and e_2 in s, we first represent each of its word w_i as a k-size (k is a hyper-parameter) word sequence which contains its left and right surrounding context words. For clarity, we note such word sequence for word w_i as WS_i in the subsequent section. For example, when we take $k = 5$, the third word w_3 in s is represented as $[w_1, w_2, w_3, w_4, w_5]$. If we consider the whole sentence s can be represented as sequence $\{[w_{pb}, w_{pb}, w_1, w_2, w_3], [w_{pb}, w_1, w_2, w_3, w_4], ..., [w_{n-2}, w_{n-1}, w_n, w_{pe}, w_{pe}]\}$ where w_{pb} and w_{pe} represent the beginning and ending padding words respectively.

For each word, we extract its features based on its word sequence representation. From Fig. 2 we can see that there are two types of features for each word: one is the syntactic parsing tree embedding feature for its k-size word sequence representation (marked as *TreeVec* in Fig. 2), the other is its lexical level features (marked as *OthVec* in Fig. 2). Finally, these two kinds of features are concatenated into a unified feature representation for the considering word. Besides, we also extract the syntactic parsing tree embedding feature between the two marked entities in the input sentence (marked as *EntityTreeVec* in Fig. 2).

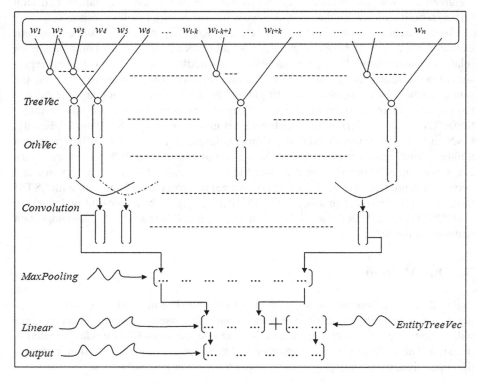

Fig. 2. Architecture of our method

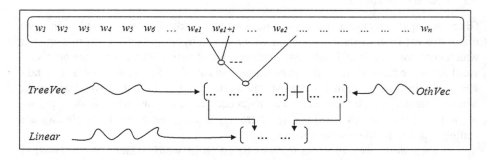

Fig. 3. The framework used for extracting *EntityTreeVec* feature demonstrated in Fig. 2

3.1.1 Syntactic *Parsing Tree Embedding Features*

Syntactic tree structure can carry more semantic and syntactic information compared with character, word, or phrase. Inspired by this fact, we think it is necessary to combine

the syntactic tree structure features extracted from the original input sentence into relation classification decision. Specifically, for each WS_i, we first obtain its syntactic parsing
tree result by using the Stanford Parsing Toolkit. Then, each syntactic tree is mapped to
a vector of dimension dim_{tree} (a hyper-parameter). In our method, this vector is initialized
using the method proposed by Socher et al. (2013a, b). In Socher et al. (2013a, b), they
extract a vector representation for each node in a syntactic parsing tree. Different from
them, in our method, we assign a vector representation for the whole structure of the
parsed tree structure. Taking the WS_i "*It also uses the intranet*" for word "*uses*" as an
example, our method first extracts its syntactic parsing tree result as shown in Fig. 4.
Then removing leaf nodes, the whole syntactic structure of the remained parsed tree is
extracted, which is "*(ROOT (S (NP (PRP)) (VP (ADVP (RB)) (VBZ) (NP (DT)
(NN)))))*". Finally, we take the extracted tree structure as input and feed it into the system
proposed by Socher et al. (2013a, b) to get its vector representation.

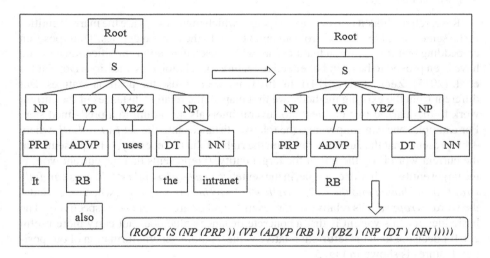

Fig. 4. An example of our method for extracting the syntactic parsing tree features

3.1.2 Lexical Level Features

There are three kinds of lexical level features used in our method: word embedding
features, POS features, and position features. Final *OthVec* is the concatenation of these
three kinds of features.

- Word Embedding Features

Word embedding is a kind of word representation method and is widely used in deep
neural networks. It converts a word into a real-valued vector representation to capture
the rich syntactic and semantic information possessed by the word. Generally, a word
embedding is a $d^w * |V|$ real-valued matrix, where d^w is the vector's dimension (a hyper-
parameter) and $|V|$ is the vocabulary size. In this matrix, each column vector corresponds
to a word representation. In our method, for each word w_i, we extract a smaller w-size

($w \leqslant |WS_i|$ and is a hyper-parameter) context window from its WS_i, and the word embedding for each word in this w-size context window is concatenated together to form the word embedding feature for w_i.

- POS Features

Part-of-speech (POS for short) features are widely used in traditional feature based relation classification methods. It is convincing that POS can carry some useful cues for relation decision. For example, a conjunction "*because*" would be more useful than a possessive pronoun "*his*" for decision. So we think it is necessary to use the POS features for each word in an input sentence. In this paper, each POS tag will be mapped to a real-valued vector of dimension dim_{pos} (a hyper-parameter). And this vector is randomly initialized.

- Position Features

Researchers think it is necessary to specify which input tokens are the marked entities in the sentence or how far it is from an input token to the marked entities. Thus, position embedding features (position features) are widely used in relation classification task and have been proved to be a kind of effective feature for relation classification (Dos Santos et al. (2015); Zeng et al. (2014)). In this work we also use the position features. But different from the existing methods, we first map all the relative distances of the current work to the marked two entities into several intervals according to a predefined step. For example, one can simply map the relative distances into the set $\{+1, -1, 0\}$, where $+1$ represents that the current word is on the right of the target entity, -1 represents that the current word is on the left of the target entity, 0 represents that the current word is the target entity itself. For example, in the sentence "*The <e1> child </e1> was carefully wrapped and bound into the <e2> cradle </e2> by means of a cord*", when we consider the word "*wrapped*", its relative distances to *e1* and *e2* are $+1$ and -1 respectively. The basic idea for this variant is that a concrete distance value does not carry more useful cue for decision than the left/right relative value. A more general formation of our position features is shown in Fig. 5.

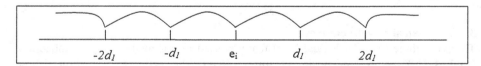

Fig. 5. Position features used in our method

In Fig. 5, a predefined step value d_l (a hyper-parameter) segments the whole coordinate system into m (a hyper-parameter) intervals. And these intervals are used to mark the relative distance from a considering word to a target entity. Each interval will be mapped into a real valued vector of dimension dim_{dst} (a hyper-parameter) and this vector is randomly initialized.

3.1.3   Entities Parsing Tree Embedding Features

Following previous researchers' idea, we think that the words between the two marked entities could provide more useful cues for relation decision and should be paid more attention. So in this paper, we take an attention scheme to enhance the features extracted from the words between the two marked entities. The enhanced feature is called *EntityTreeVec* demonstrated in Fig. 2 and its extraction method is shown in Fig. 3.

3.2   Convolution Operation

The convolutional transformation is a kind of linear transformation and is expected to extract some useful features from the aspect of short phrases. Given a represented sentence x_i, the convolutional transformation applies a matrix-vector operation to each of its word sequence feature x_i. The matrix used is often called as a filter that is noted as $Mtr_{con} \in R^{h1*dws}$ in this paper. Here h_1 (a hyper-parameter) is the size of hidden units and $d_{ws} = |tv| + |ov|$ is the dimension of feature vector for word sequence. The convolutional transformation operation can be denoted as following formula where b is a bias term.

$$C_i = Mtr_{con} * x_i + b \tag{1}$$

Mtr_{con} will be applied to each word sequence x_i in the input sentence x_s.

3.3   Max-pooling Operation

After the convolutional transformation, a max-pooling operation over the sentence length n_i is often applied to capture the most useful local features produced by the convolutional transformation operation. The max-pooling process can be written as formula (2).

$$p_i = max_n C(i, n) \ 0 \le i \le h_1 \tag{2}$$

3.4   Linear Transformation

After the max-pooling operation, its output vector p is fed to a linear transformation layer to perform affine transformation over the input as following formula (3).

$$f_1 = tanh(Mtr^{l_1} \times p + b^{l_1}) \tag{3}$$

Where $Mtr^{l_1} \in R^{h_2 \times h_1}$ is the linear transformation matrix where h_2 (a hyper-parameter) is the size of hidden units in this layer l_1 and b^{l_1} is a bias term.

After *EntityTreeVec* is extracted, we will obtain a feature vector m, another linear transformation layer will perform affine transformation over it with formula (4).

$$f_2 = tanh(Mtr^{l_2} \times m + b^{l_2}) \tag{4}$$

Where $Mtr^{l_2} \in R^{h_3 \times dim_e}$ is the linear transformation matrix where h_3 is the size of hidden units in this layer l_2 and dim_e is the dimension for feature vector m.

3.5 Output

After the above linear transformations, their outputs are concatenated into a unified vector $f = [f_1, f_2]$ to represent a higher level feature. Then f is fed into a linear output layer to compute the confidence score for each possible relation type. We use a *softmax* classifier to get the probability distribution y over relation labels. This process can be shown in following formula (5).

$$y = softmax(Mtr^{out} \times f + b^{out}) \tag{5}$$

Where $Mtr^{out} \in R^{h_4 \times dim_f}$ and h_4 is the number of possible relation types, dim_f is the dimension of feature vector f, and b^{out} is a bias term.

3.6 Dropout Operation

In deep neural network, over-fitting is an issue that cannot be ignored. Currently, dropout method that is proposed recently by Hinton et al. (2012) has been proved to be an effective regularization approach for alleviating over-fitting. It randomly omits a proportion (called as drop rate, denoted as d_{pro} here, a hyper-parameter) of features during training to obtain less interdependent network units, so that better performance is expected to be achieved. In our method, we dropout strategy is taken at both the feature extraction phase and the linear transformation phase. Specially, we take dropout operation on the feature x_i in formula (1), m in formula (4) and f in formula (5). Then these 3 formulas will be updated as following formulas where D denotes the dropout operator.

$$C_i = Mtr_{con} \times D_{d_{pro1}}(x_i) + b \tag{1'}$$
$$f_2 = tanh(Mtr^{l_2} \times D_{d_{pro2}}(m) + b^{l_2}) \tag{4'}$$
$$y = softmax(Mtr^{out} \times D_{d_{pro3}}(f) + b^{out}) \tag{5'}$$

3.7 Training Procedure

All of the parameters in this paper can be denoted as $\theta = (E^W, E^t, E^l, E^p, Mtr_{con}, Mtr^{l1}, Mtr^{l2}, Mtr^{out}, b, b^{l1}, b^{l2}, b^{out})$ where E^W, E^t, E^l, E^p represent the embeddings for word, syntactic tree, position, and POS respectively. And E^w is initialized by the trained embeddings provided by SENNA (Collobert et al. 2011), E^t is initialized by the method introduced in Socher et al. (2013a, b). Other embeddings, transformation matrixes and bias terms are randomly initialized. All of the parameters are tuned using the back propagation method and stochastic gradient descent (SGD) optimization technique is used during the training procedure. Formally, we try to maximize the following objective function.

$$J(\theta) = \sum_{i=1}^{N} logy_i \tag{6}$$

Where N is the total number of training samples. During training, each input sentence is considered independently, and each parameter is updated by applying the following update rule. Where γ is the learning rate.

$$\theta = \theta + \gamma \frac{\partial logy}{\partial \theta} \tag{7}$$

4 Experiments

The SemEval-2010 Task 8 dataset is used here to evaluate our method. In this dataset, there are 8000 training sentences and 2717 test sentences. Two entities that are expected to be predicted a relation are marked for each sentence. There are 9 relation types whose directions need to be considered and an extra artificial relation *"Other"* which does not need to consider the direction. There are 19 relation types totally in this relation classification task. The official evaluation metric macro-averaged F1 score (excluding *Other*) is used here and the direction is considered. All the syntactic parsing trees are generated by the Stanford Parser (Klein and Manning 2003). We apply a cross-validation procedure on the training data to select suitable hyper-parameters.

Our experiments consist of two parts: we first evaluate the contributions of different features, and then we compare our method with some state-of-the-art methods. We implement a CNN method that is similar to the one described by Zeng et al. (2014) as a baseline system. In the baseline system, we only word embedding feature is used. Then we investigate how the F1 result changes when we successively add further features to the baseline. Table 1 shows the results. Experimental results show that the performance increases greatly when the tree embedding features added, even greater than performance gain obtained when position feature, a kind of very effective feature for relation classification task which has been proved by many previous research work, is added.

Table 1. Performance of our method with different features

Our model	F1
Baseline	59.4
+ tree emb features	73.4
+ position features	69.1
+ POS features	60.2
+ *EntityTreeVec*	63.7
ensemble	82.4

We also compare our method with other state-of-the-art methods, the comparison results are shown in Table 2, where CNN refers to the method proposed by Zeng et al. (2014) and other classifier methods and their corresponding features and F1 score are borrowed directly from the paper described by Zeng et al. (2014).

Table 2. Comparison with other methods

Classifier	Additional features used	F1
CNN	WordNet	82.7
SVM	POS, prefixes, morphological, WordNet, dependency parse, Levin classed, ProBank, FrameNet, NomLex-Plus, Google n-gram, paraphrases, TextRunner	82.2
RNN	–	74.8
	POS, NER, WordNet	77.6
MVRNN	–	79.1
	POS, NER, WordNet	82.4
Proposed	parsing trees, POS	82.4

From Table 2 we can see that our method achieves similar results compared with the other methods. However, we must point out that our method is the only one that does not use WordNet feature. Considering the condition of not using WordNet, our method actually achieves the best results. Our experimental results show the effectiveness of embedding the syntactic parsing tree into a CNN architecture for relation classification.

5 Conclusions

In this paper, we embed the syntactic parsing tree structures into a CNN architecture for relation classification. The main contributions of this work are as following.

1. We use the syntactic parsing tree structures in a deeper manner compared with most of current researchers. In our method, each k-size word sequence is parsed into a tree structure and its embedding features are extracted and fed into a CNN model.
2. In our method, we don't use any external linguistic resources like WordNet. Experimental results show that the state-of-the-art performance is achieved with our method at the condition of without using external resources.

On the other hand, there are still some other issues needed to be further investigated. First, in our method, the parsing tree structures are extracted from some short text. This will introduce lots of noise tree structures because of the fact that the source short text may not carry a complete semantic meaning itself. Second, the tree embedding initializing method is used by the method described by Socher et al. (2013a, b). This method doesn't aim to generate tree embedding at all, so the initialed tree embedding would not be suitable for the task here. All of these issues will be further investigated in our further work.

Acknowledgements. This work is supported by the National Natural Science Foundation of China (NSFC No. 61572120, 61300097 and 61432013). We thank all anonymous reviewers for their constructive comments.

References

Collobert, R., Weston, J., Bottou, L., Karlen, M., Kavukcuoglu, K., Kuksa, P.: Natural language processing (almost) from scratch. J. Mach. Learn. Res. **12**, 2493–2537 (2011)

Wang, L., Cao, Z., de Melo, G., Liu, Z.: Relation classification via multi-level attention CNNs. In: Proceedings of the 54th Annual Meeting of the Association for Computational Linguistics, pp. 1298–1307 (2016)

Cai, R., Zhang, X., Wang, V.: Bidirectional recurrent convolutional neural network for relation classification. In: Proceedings of the 54th Annual Meeting of the Association for Computational Linguistics, pp. 756–765 (2016)

Xu, K., Feng, Y., Huang, S., Zhao, D.: Semantic relation classification via convolutional neural networks with simple negative sampling. In: Proceedings of 2015 Conference on Empirical Methods in Natural Language Processing, pp. 536–540 (2015a)

Xu, Y., Mou, L., Li, G., Chen, Y., Peng, H., Jin, Z.: Classifying relations via long short term memory networks along shortest dependency paths. In: Proceedings of the 2015 Conference on Empirical Methods in Natural Language Processing, pp. 1785–1794 (2015b)

Zeng, D., Liu, K., Lai, S., Zhou, G., Zhao, J.: Relation classification via convolutional deep neural network. In: Proceedings of the 25th International Conference on Computational Linguistics, pp. 2335–2344 (2014)

Socher, R., Bauer, J., Manning, C.D., Ng, A.Y.: Parsing with compositional vector grammars. In: Proceedings of the 51th Annual Meeting of the Association for Computational Linguistics, pp. 455–465 (2013a)

Socher, R., Perelygin, A., Wu, J.Y., Chuang, J., Manning, C.D., Ng, A.Y., Potts, C.: Recursive deep models for semantic compositionality over a sentiment treebank. In: Proceedings of the 2013 Conference on Empirical Methods in Natural Language Processing, pp. 1631–1642 (2013b)

dos Santos, C.N., Xiang, B., Zhou, B.: Classifying relations by ranking with convolutional neural networks. In: Proceedings of the 53rd Annual Meeting of the Association for Computational Linguistics, pp. 626–634 (2015)

Liu, Y., Wei, F., Li, S., Ji, H., Zhou, M., Wang, H.: A dependency-based neural network for relation classification. In: Proceedings of the 53rd Annual Meeting of the Association for Computational Linguistics, pp. 285–290 (2015)

Vu, N.T., Adel, H., Gupta, P., Schutze, H.: Combining recurrent and convolutional neural networks for relation classification. In: Proceedings of NAACL-HLT 2016, pp. 534–539 (2015)

Bach, N., Badaskar, S.: A review of relation extraction. Literature review for Language and Statistics H (2007)

Hashimoto, K., Miwa, M., Tsuruoka, Y., Chikayama, T.: Simple customization of recursive neural networks for semantic relation classification. In: Proceedings of the 2013 Conference on Empirical Methods in Natural Language Processing, pp. 1372–1376 (2013)

Socher, R., Huval, B., Manning, C.D., Ng, A.Y.: Semantic compositionality through recursive matrix-vector spaces. In: Proceedings of the 2012 Joint Conference on EMNLP and Computational Natural Language Learning, pp. 1201–1211 (2012)

Klein, D., Manning, C.D.: Accurate unlexicalized parsing. In: Proceedings of 2003 ACL, pp. 423–430 (2003)

Petrov, S., Klein, D.: Improved inference for unlexicalized parsing. In: Proceedings of NAACL HLT 2007, pp. 404–411 (2007)

Hinton, G.E., Srivastava, N., Krizhevsky, A., Sutskever, I., Salakhutdinov, R.R.: Improving neural networks by preventing co-adaptation of feature detectors. arXiv preprint arXiv:1207.0580 (2012)

Ebrahimi, J., Dou, D.: Chain based rnn for relation classification. In: Proceedings of the 2015 Conference of the North American Chapter of the Association for Computational Linguistics: Human Language Technologies, pp. 1244–1249 (2015)

Zhang, D., Wang, D.: Relation classification via recurrent neural network. arXiv preprint arXiv: 1508.01006 (2015)

Tracking Topic Trends for Short Texts

Liyan He$^{(\boxtimes)}$, Yajun Du, and Yongtao Ye

School of Computer and Software Engineering, Xihua University,
Chengdu 610039, China
15608090757@163.com

Abstract. It is a critical task to infer discriminative and coherent topics from short texts. Furthermore, people not only want to know what kinds of topics can be extract from these short texts, but also desire to obtain the temporal dynamic evolution of these topics. In this paper, we present a novel model for short texts, referred as topic trend detection (TTD) model. Based on an optimized topic model we proposed, TTD model derives more typical terms and itemsets to represent topics of short texts and improves the coherence of topic representations. Ultimately, we extend the topic itemsets obtained from the optimized topic model by word embeddings to detect topic trends. Through extensive experiments on several real-world short text collections in Sina Microblog, the result demonstrate our method achieves comparable topic representations than state-of-the-art models, measured by topic coherence, and then show its application in identifying topic trends in Sina Microblog.

Keywords: Topic model · Short text · Word embedding
Trend detection

1 Introduction

Researchers have shown an increased interest recently in analysing user intent, hot topics and social behaviors among these social platforms such as Twitter, Facebook and Microblog. Topic models, as a class of newly developed machine learning techniques, are employed to identify the underlying topics from text corpus. These topic models such as LDA [1] and pLSA [2], partly consider the semantic relatedness between words by incorporate topic information into a multinomial distribution over documents and words. Statistical techniques (e.g., Gibbs sampling) are then employed to identify the underlying topic distribution of each document as well as word distribution of each topic, based on the high-order word co-occurrence patterns [3,4]. However, these original topic models encounter a large performance degradation over short texts because of limited word co-occurrence information in short texts. Based on these models' variants and others' emerging technologies, such as neural network language models [5,6], several major heuristic strategies have been adopted to solve the data sparsity problem in short texts. Some researchers aim to aggregate a subset of short

© Springer Nature Singapore Pte Ltd. 2017
J. Li et al. (Eds.): CCKS 2017, CCIS 784, pp. 117–128, 2017.
https://doi.org/10.1007/978-981-10-7359-5_12

texts to form a long texts. And then topic models are applied over these long texts [7]. Other ingenious methods, such as MA-LDA [8] and MB-LDA [9], take some features into consideration and incorporate these features into LDA. It can outperforms the baseline of LDA empirically. Specifically, with the recent development in neural network techniques, Word embedding, first introduced in [10], have been successfully applied in language models and many NLP tasks. The combination of topic model and neural network language model prove to be a feasible way to enhance semantic relatedness [11].

Nowadays, people not only want to know what kind of topics can be extract from these text contents but also desire to gain the temporal dynamics of these topics. This calls for a more effective model that can track the evolution of trends over time in the analysis of user behavioral data. Many models have been proposed to incorporate temporal dynamics into topic models. The typical example in this category is Dynamic Topic Model (DTM), proposed by Blei [12]. Another model called Trend Detection Model (TDM), introduces a latent trend class variable into each document. However, a limitation of these methods is that word co-occurrence information that can be captured by these models because of data sparsity.

In this paper, we propose a novel model named topic trends detection (TTD) model. It contains an optimized topic model, which incorporates a term weighting pattern into a standard LDA model. Unlike existing models in which just incorporate time parameters and observed variables into Probabilistic generative model such as DMM [13–15], we combine our topic model with neural network language model in an unsupervised learning fashion which enables us to effectively and accurately detect the topic trends. In addition, the background knowledge about word semantic relatedness learned from a series of relevant short texts can be easily exploited to improve topic modeling by word embeddings. Through extensive experiments on several short text collections, we show our model's effectiveness and accurateness over non-trivial baselines. The main contributions of this paper are summarized as follows:

(1) We propose an effective and feasible model which can not only extract latent topics over short texts correctly but also track temporal dynamic trends of these topics.
(2) We firstly present a newly topic model to generate more discriminative topics from aggregates of short texts accurately.
(3) Demonstrate the effectiveness of the model as compared to baseline models by experimenting on several microblog text collections.

The remainder of this article is organised as follows. In Sect. 2, we review some previous work on extracting topics over short texts and tracking the topic trends over time. In Sect. 3, we formalise our optimized topic model and TTD models. Our main experiments and results are presented in Sect. 4. The conclusions are discussed in Sect. 5.

2   Related Work

In this section, we review three directions of related work. First, we summarize several classical topic models which try to infer topics for short texts. Second, we discuss the ingenious strategy of advanced models with word embeddings.

Topic models posit that each document is expressed as a mixture of topics, which are designed for categorical data. These topic proportions of standard topic model such as LDA [1], are drawn once per document, and the topics are shared across the corpus. This model is widely used to extract latent topic from text corpus. Another variant based on LDA called DMM shows that each short text is sampled from a single topic, known as mixture of unigrams or Dirichlet Multinomial Mixture (DMM) [13–15]. Blei et al. propose a correlated topic model (CTM) [16] that treats each document's topic assignment as a multinomial random variable drawn from logistic normal prior. Yan et al. propose a novel way for modeling topics in short texts, referred as biterm topic model (BTM) [17]. It learn the topics by directly modeling the generation of word co-occurrence patterns. However, there is still a problem that these methods can not fully capture the semantic relatedness between the words. With the recent development in neural network techniques, word embedding have contributed incremental improvements in researching word semantic relatedness [14,18,19].

Word embeddings, first introduced in [10], have been shown to capture lexico-semantic regularities in language: Words with similar syntactic and semantic properties are found to be close to each other in the embedding space [20,21]. According to the distributional hypothesis [22], words occurring in similar contexts tend to have similar meaning. Liu et al. propose a model called Topical Word Embeddings (TWE) [23], which employ latent topic models to assign topics for each word in the text corpus, and learn topical word embeddings based on both words and their topics. Multi-prototype vector space models [24] were proposed to cluster contexts of a word into groups, then generate a distinct prototype vector for each cluster.

3   Topic Trend Detection (TTD) Model

TTD model is built to track topic trends for short texts based on a novel topic model and a effective neural network language model. Given a short document collections belong to a certain event, an optimized topic model we proposed extracts several topic terms or itemsets according to its weighting pattern and conditional probabilities. As shown in Fig. 1, the word embeddings utilized in TTD is pre-learned using the state-of-the-art word embedding techniques from current date collections. Next, we present the details of the proposed model TTD.

3.1   Pre-process Stage

As is known to us, the standard LDA model can't perform well over short texts due to the data sparsity. We consider it from two aspects. The first intuitive

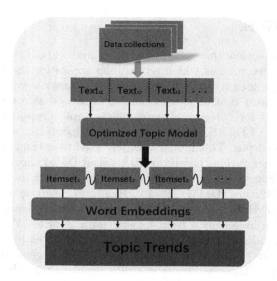

Fig. 1. Topic Trend Detection (TTD) model overview

solution to this problem is pooling method: merging related short texts together and presenting them as a single document to stand topic model. In this paper, we put up with a novel short text-pooling scheme that we aggregate short texts in a certain period of time to form a longer pseudo-document. Our topic model is then applied over these pseudo-documents. Since this method specifically designed for short text contents, it will eliminate the additional unavailable metadata absolutely and enhance the correlation between these documents. On the other hand, we restrict the document-topic distribution like DMM model [13], such that each short text is sampled from a single topic.

3.2   An Optimized Topic Model

For most LDA based topic models, the words with high probabilities in topic's word distributions are usually chosen to represent topics. There is a common phenomenon that several different topics usually have similar topic-word distribution. That means, these words most likely represent general concepts or common concepts of the topics and cannot distinctively represent the topics. In the following work of detecting topic trends, the same words will lead to the same related words because the high dimensional representation of a certain word is fixed. In addition, the words in topic representations generated by LDA are usually single words. These single words provide too limited information about the relationships between the words and too limited semantic meaning to make the topics understandable. Accordingly, we extend LDA model by incorporating the weighting pattern to generate a distinctive topic word or itemset. In this paper, we exploit TF-IDF weight as a part of weighting pattern for our model.

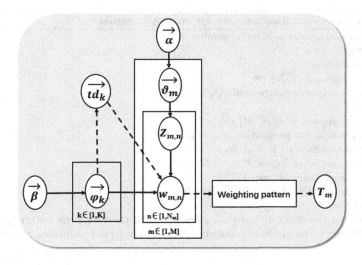

Fig. 2. A generative graphical topic model

Shown in Fig. 2, the proposed topic model extends the Latent Dirichlet Allocation (LDA) model by incorporating the weighting pattern. The details of Generation process for our topic model is described in Algorithm 1. M is the total number of documents, K is the number of latent topics, α, β is Dirichlet parameters, φ_k is word distribution for topic k, ϑ_m is topic distribution for document m, N_m is the length of document m, $Z_{m,n}$ is topic index of n_{th} word in document m, $\omega_{m,n}$ is the nth word of mth document. For the topical document, the word feature distribution over topic k, denoted as td_k, is generated based on their tf-idf scores, which are calculated by (1). $td_{i,j}$ is the jth term of ith document. $tf(td_{i,j})$ is the frequency of term $td_{i,j}$ in the ith document, m_i is the ith document that consists of the words for a particular topic, $|m_i|$ is the count of terms in m_i, $N(td_{i,j})$ is the count of $td_{i,j}$ appearing in m_i. Inverse document frequency (idf) reflects the popularity of term $td_{i,j}$ across topical documents in D_{topic}, where M is the total number of topical documents and $df(td_{i,j})$ is the document frequency. Therefore, a term has high tf-idf weighting indicates high term frequency but low overall document frequency.

$$tfidf(td_{i,j}) = tf(td_{i,j}) \times idf(td_{i,j}) = \frac{N(td_{i,j})}{|m_i|} \times \log \frac{M+1}{df(td_{i,j})} \qquad (1)$$

The hidden variables in the generative process can be approximated by applying Gibbs sampling. Following the approach in [1], word i has been assigned to a topic k is sampled for each document in every iteration according to the following conditional distribution:

$$p(z_i = k | \boldsymbol{z}_{\neg i}, \boldsymbol{w}) \propto \frac{n_{k,\neg i}^{(t)} + \beta_t}{\sum_{t=1}^{V} n_{k,\neg i} + \beta_t} \frac{n_{m,\neg i}^{(k)} + \alpha_k}{\sum_{k=1}^{K} n_{m,\neg i} + \alpha_k} \qquad (2)$$

Algorithm 1. Generation process for our topic model

Input: hyperparameter α,β,topic number K
Output: topic word T_m
  1: \\ Initialisation
  2: **for** all topics $k \in [1, K]$ **do**
  3:      sample mixture components $\varphi_k \sim Dir(\beta)$
  4: **end for**
  5: **for** all documents $m \in [1, M]$ **do**
  6:      sample mixture proportion $\vartheta_m \sim Dir(\alpha)$
  7:      **for** all words $n \in [1, N_m]$ **do**
  8:          sample topic index $Z_{m,n} \sim Mult(\vartheta_m)$
  9:          sample term for word $\omega_{m,n} \sim Mult(\vartheta_{z_{m,n}})$
 10:          calculate TF-IDF word weight in topic k
 11:      **end for**
         Adjust the order of word $\omega_{m,n}$ according to the weighting pattern
         Choose topic words of a document according to the weighting pattern
 12: **end for**

$$\varphi_{k,t} = \frac{n_k^{(t)} + \beta_t}{\sum_{t=1}^{V} n_k^t + \beta_t} \tag{3}$$

$$\vartheta_{m,k} = \frac{n_m^{(k)} + \alpha_k}{\sum_{k=1}^{K} n_m^k + \alpha_k} \tag{4}$$

In (2), Symbol $\neg i$ means that the current counting i is excluded from the total counting. α and β Dirichlet priors, $n_{k,\neg i}^{(t)}$ is the number of times that word w is assigned to topic k and exclude the current counting. V represents the vocabulary size and K is the number of pre-defined latent topics. $n_{m,\neg i}^{(t)}$ is the number of times that topic k is assigned to topic document m and exclude the current counting. The first ratio expresses the probability of word i under topic k as (3), and the second ratio expresses the probability of topic k in document m as (4). Specially, these counts are the only information necessary for computing the full conditional distribution, allowing the algorithm to be implemented efficiently by caching the relatively small set of nonzero counts. The biggest difference between the traditional topic model and the optimized topic model we proposed is that our model is built upon the weighting pattern. For each document, we get the word distribution over topic k. And then, we sort the words belong to each topic in TF-IDF value of these words as Table 1. The basic idea of the proposed method is to use the frequent itemsets generated from each document D_i to represent topic k. For a given minimal support threshold ζ, and itemset w in D_i is frequent if $F(w) >= \zeta$, where $F(w)$ is the frequent of w which is the number of itemsets in D_i that contain w. For example of D_2, set a minimal support threshold $\zeta = 2$, all the frequent itemsets generated from D_2 are given in Table 2. We can notice that $\{w_1\}$, $\{w_8\}$ and $\{w_1, w_8\}$ are the dominant topics for Document 2. Comparing with the term based topic representation, itemsets represent the correlative words that carry more concrete and identifiable meaning. For example, "Houston

Table 1. The topic-word distribution of each document based on TF-IDF value

D_1		D_2		D_3	
Topic	Topic-word	Topic	Topic-word	Topic	Topic-word
1	W_1,W_2,W_3	1	W_1,W_8,W_9	1	W_7,W_{10}
2	W_2,W_3,W_5	2	W_1,W_7,W_8	2	W_1,W_{11},W_{12}
3	W_1,W_3,W_5,W_7	3	W_2,W_3,W_6	3	W_4,W_7,W_9,W_{11}
4	W_2,W_3	4	W_1,W_5,W_8,W_9	4	W_4,W_6,W_8

Table 2. Statistics on topic itemsets of Document 2

Itemsets	$F(w)$
$\{w_1\}, \{w_8\}, \{w_1, w_8\}$	3
$\{w_9\}, \{w_1, w_9\}$	2

Rockets" is more concrete than just one word "Houston" or "Rockets". In this paper, we treat itemsets as the dominant topics for the aggregates of short texts. For tracking trends of complete events, we aggregate the microblog texts of one event in a time period to a document and then obtain the dominant topics from Topic model. Finally, we extract detailed information using word embeddings.

3.3 Auxiliary Word Embeddings

Although our topic model can extract discriminative terms in short texts to some extent, this prior preference for semantic coherence is not encoded in the model, and any such observation of semantic coherence found in the inferred topic distributions is, in some sense, accidental. In other words, Given two words having strong semantically relatedness but rarely co-occurring in a short text corpus, these models can not fully capture the semantic relatedness between the two words. At end of our TTD model, we utilize neural network language model for exploiting the contiguity of semantically similar words in the embedding space. In the experiment, we use Google's Word2Vec toolkit with Skip-gram algorithm because Skip-gram algorithm can perform well on small date collections and catch the distinctive words to represent the trends.

4   Experiments

4.1   Datasets

In this section, we conduct extensive experiments to evaluate the proposed TTD model against the baseline models. The text content in Sina Microblog is short because of the platform constraints. In this circumstance, we can regard microblog content as short texts. We purposely collect the reliable text corpora including 17919 sina microblog texts from an authoritative website[1]. And we aggregate these texts according to different events to several documents for training word embeddings. These microblog contents belong to several hot events.

Table 3. Statistics on the several event datasets

#Event	#Docs	#W/D	Vocabulary
#BaiHe Bai infidelity#	#4138	#8.64	3334
#Yan Liu with bridesmaid#	#5438	#10.76	4980
#HeShuo Hotel#	#2865	#9.77	2267
#Sa De#	#5289	#12.34	5783
#The retire of Ray Allen#	#189	#9.23	130

Statistics on the datasets after preprocessing is reported in Table 3:[2] #Docs: the total number of documents; #W/D: average number of words per document; Vocabulary: the Vocabulary capacity.

4.2 Experimental Setup

We removed non-alphabetic characters and stop-words found in our WF-LDA, and we also removed words appearing less than 5 times in event aggregates for Word2Vec toolkit. We use Google's Word2Vec toolkit with Skip-gram algorithm [5] and if a word has no embedding, the word is considered as having no word semantic relatedness knowledge. As mentioned in Sect. 3.2, we set minimal support $\zeta = 3$ and choose the topic itemsets from top 10 words with high probabilities in topics' word distributions. For all the methods in comparison, the hyper-parameter used in baseline LDA and DMM models was set to 0.01. We set the hyper-parameter $\alpha = 0.1$, as this can improve performance relative to the standard setting $\alpha = \frac{50}{T}$. Since microblog texts have the characteristic of immediacy, the release time of every microblog is considered as a certain point in time among the event evolution. We apply aggregate pattern to all of models for the fairness of comparison.

4.3 Evaluation by Topic Coherence

The topics generated by each model are evaluated by the topic coherence metric. Conventionally, topic models are evaluated by perplexity. However, there is a limitation that perplexity can not reflect the semantic coherence of a topic. Accordingly, it can sometimes be contrary to human judgments. Topic coherence

---

[1] http://www.zhiweidata.com/.

[2] In the following paper, the event name on microblog will be replaced by English to avoid the Chinese problems in Tex.

measures the extent that the most probable words of a topic tend to co-occur together within the same documents. It proved to be a better metric to assess topic quality.

Here, we use the PMI-Score proposed in [25] to calculate topic coherence. PMI has been studied extensively in the context of collocation extraction. Given a topic k and its top T words with highest probabilities $(w_1, \cdots, w_T)$, the PMI-Score of k is:

$$PMI(k) = \frac{2}{T(T-1)} \sum_{1 \leq i \leq j \leq T} \log \frac{p(w_i, w_j)}{p(w_i)p(w_j)} \tag{5}$$

In (5), $p(w_i)$ refers to the probability that word w_i appears in a document, and $p(w_i, w_j)$ means the probability that words w_i and w_j appear in the same document. The overall topic coherence for each model is the averaged PMI-Score over all learnt topics. A higher topic coherence indicates the better learnt topics. Note that it need an external corpus, we use 3 hundred thousand Sina Microblog texts for our experimental datasets since microblog language has its own distinctive features. Figure 3 shows PMI-Score computed for the baseline LDA, DMM and TTD models on the microblog dataset with number of top words per topic $T = \{5, 10, 15\}$ and number of topics $K = \{30, 50, 70\}$, respectively. As we can see, our model has a better performance in topic coherence across all settings. That is because we incorporate a novel weighting pattern into the standard topic model. It can detect more representative words. DMM model is the second best model in most cases, and always outperforms LDA. Since it is a specific model for short texts which assumes each short document has only one topic. Compared with other methods, baseline LDA model has a poor performance in topic coherence for the limited word co-occurrence information, which impedes the generation of discriminative document-topic distributions.

Figure 4 presents the trends of PMI score in different number of topics vary from 30 to 70. We make the following observations. As selecting more Top-N words, the PMI score by every model shows descending trend in the same number of Topics. The reason is that choosing more words means the implication of more relevant topics. From (5) we can see, it increases more word pairs and reduce the scores of PMI. For learning word embeddings and evaluating topic coherence, we used jieba, a python package segmentation tool, to segment the

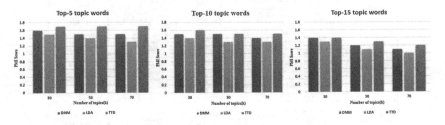

Fig. 3. Topic coherence on microblog dataset

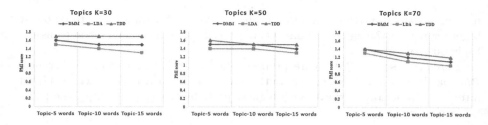

Fig. 4. Variation trends of choosing different topics

words in the microblog datasets. As is known to us, microblog language is completely informal. The accuracy in Chinese word segmentation might cause some performance variations.

4.4 Trend Detection in Microblog

Ultimately, we are interested in utilizing our method over microblog events to detect emerging trends in real-time. Since Sina Microblog has the property of immediacy, we treat each day as an unit to analysis the trend evolution by using TTD. As shown in Table 4, we took the $R(M)$ as the ratio of the amount of microblog texts that day to the amount of microblog texts of the whole

Table 4. Several hot event trends in microblog

#event#	Date	R(M)	Topic words	Related words
#白百合出轨#	2017.4.11	0.01	白 百合	歌手 风云 期待 称东
	2017.4.12	0.55	出轨	白百合 陈羽凡 娱乐圈 羽泉 卓伟
	2017.4.13	0.19	出轨	陈羽凡 离婚 左拥右抱 美女 视频
	2017.4.14	0.11	出轨 陈羽凡	白百合 人品 生活 女人 离婚
	2017.4.15	0.03	陈羽凡 发文	发文 牵手 女人 毁 亡
	2017.4.16	0.1	陈羽凡	声明 退出 视频 娱乐圈 白百合
	2017.4.17	0.01	离婚	好看 推荐 爆笑 陈羽凡 网友
#柳岩伴娘#	2016.3.30	0.02	包贝尔 婚礼	伴郎 柳岩 婚庆 补办 贾玲
	2016.3.31	0.31	柳岩 婚礼	穿 女性 视频 开玩笑 回应
	2016.4.1	0.14	伴娘	包贝尔 伴郎 捉弄 杜海涛 扔
	2016.4.2	0.28	婚礼 网友	恶意 风波 哽咽 致歉 回应
	2016.4.3	0.06	网友 包文婧	指责 婚礼 事件 包贝尔 致歉
	2016.4.4	0.04	捉弄 柳岩	伴娘 婚礼 韩庚 李紫宰 清明节
	2016.4.5	0.03	太阳 公安 千玺	闹 公安 超能量 易 捉弄
	2016.4.6	0.06	豪姿 伴娘	禁止 婚闹 协议 闺蜜 婆
	2016.4.7	0.05	闹婚 伴娘	王女士 绅士 上树上线 风度 男性
	2016.4.8	0.01	绅士 风度	知识 崔始源 雷建华 蕾 题
#和硕酒店#	2016.4.5	0.06	酒店 弯弯	硕 集团 地上 强奸 拖拽
	2016.4.6	0.77	女生 酒店	酒店 发表 事件 日 弯弯
	2016.4.7	0.09	酒店 硕	声明 女生 强行 拖拽 视频
	2016.4.8	0.08	酒店 涉案	北京警方 事件 采访 男子 政府
#萨德事件#	2017.2.26	0.01	韩国	恩凯 乐 支持 买 乐天 超市
	2017.2.27	0.07	中国 韩国	乐天 萨德 月 部署 恭父
	2017.2.28	0.18	抵制	中方 态度 愤怒 市场 企业
	2017.3.1	0.16	华 企业	乐天 抵制 美国 新闻 超市
	2017.3.2	0.23	抵制 卫龙	产品 民众 官方 辣条 厉害
	2017.3.3	0.09	乐天 中国	超市 企业 韩国 抵制 全文
	2017.3.4	0.06	萨德 乐天	中国 反对 政府 话题 部署
	2017.3.5	0.07	韩国 抵制	购物 朝鲜 超市 爱国 集团
	2017.3.6	0.11	中国 乐天	冯特 免税店 涨 部署 民族
#雷阿伦退役#	2016.11.1	0.45	雷 阿伦 退役	论坛 黄金 三分球 命中 正式
	2016.11.2	0.55	正式 传奇	黄金 一代 职业生涯 凯尔特人 热火

event to discover the level of public concerns. Topic words are also the itemsets obtained from our model. Related words indicate the top 5 words obtained from Google's Word2Vec toolkit with Skip-gram algorithm that have the highest similarity compare to topic itemsets. From above results, note that the amount of Microblog collections in the second day usually has a higher proportion. When a hot event emerged firstly, the Microblog amount should increase quickly in a short time. That means more people begin to concentrate on a specialized topic and discuss about it. And then, Microblog amount in next days decreases in some extent, which indicates that people have acquired the detailed information and they reduce the overall concern about the current event unless another relevant event appears. If a hot event is inconclusive in the near future, it will keep relatively higher attention in the end like #Sa De#. Notice that several results in details are relatively irrational especially in #HeShuo Hotel# and #The Retire of Ray Allen#. That because the Microblog text collections of #HeShuo Hotel# are not according to the event occurrence time but the time after. For #The Retire of Ray Allen#, as we can see, the problem of data sparsity affects the results to some extent.

5   Conclusion and Future Work

Instead of extracting a individual single word as a topic from a document, an optimized model we proposed incorporates word features and LDA model and chooses discriminative and semantic rich representations for modeling the topics in a aggregation of short texts. Finally we present its effectiveness over baseline models.

However, there is still room to improve our model in the future. For example, we would like to consider other forms of topic models with other word features to enable more effective learning. Moreover, it is interesting to test and verify the effectiveness of using other word embedding techniques like Glove [6].

Acknowledgement. This work is funded by the National Natural Science Foundation of China under Grant No. 61472329, No. 61532009 and the Innovation Fund of Xihua University. We would like to thank the anonymous reviewers for their helpful comments.

References

1. Blei, D.M., Ng, A.Y., Jordan, M.I.: Latent Dirichlet allocation. J. Mach. Learn. Res. **3**, 993–1022 (2003)
2. Hofmann, T.: Probabilistic latent semantic indexing. In: SIGIR (1999)
3. Li, C., Wang, H., Zhang, Z., Sun, A., Ma, Z.: Topic modeling for short texts with auxiliary word embeddings. In: SIGIR (2016)
4. Porteous, I., Newman, D., Ihler, A., Asuncion, A., Smyth, P., Welling, M.: Fast collapsed Gibbs sampling for latent Dirichlet allocation. In: SIGKDD (2008)
5. Mikolov, T., Chen, K., Corrada, G., Dean, J.: Efficient estimation of word representations in vector space. arXiv preprint arXiv:1301.3781 (2013)

6. Pennington, J., Socher, R., Manning, C.D.: Glove: global vectors for word representation. In: EMNLP (2014)
7. Jin, O., Liu, N.N., Zhao, K., Yu, Y., Yang, Q.: Transferring topical knowledge from auxiliary long texts for short text clustering. In: Proceedings of the 20th ACM International Conference on Information and Knowledge Management, pp. 775–784 (2011)
8. Wang, J., Li, L., Tan, F., Zhu, Y., Feng, W.: Detecting hotspot information using multi-attribute based topic model. Plos One **10**(10), e0140539 (2015)
9. Zhang, C., Sun, J.: Large scale microblog mining using distributed MB-LDA. In: WWW Companion (2012)
10. Rumelhar, D.E., Hinton, G.E., Williams, R.J.: Learning representations by back propagating errors. **323**(6088), 533–536 (1988). MIT Press
11. Nguyen, D.Q., Billingsley, R., Du, L., Johnson, M.: Improving topic models with latent feature word representations. TACL **3**, 299–313 (2015)
12. Blei, D.M., Lafferty, J.D.: Dynamic topic models. In: Proceedings of the 23rd International Conference on Machine Learning (ICML) (2006)
13. Nigam, K., MacCallum, A.K., Thrun, S., Mitchell, T.: Text classification from labeled and unlabeled documents using EM. Mach. Learn. **39**, 103–134 (2000)
14. Zhao, W.X., Jiang, J., Weng, J., He, J., Lim, E.-P., Yan, H., Li, X.: Comparing twitter and traditional media using topic models. In: Clough, P., Foley, C., Gurrin, C., Jones, G.J.F., Kraaij, W., Lee, H., Mudoch, V. (eds.) ECIR 2011. LNCS, vol. 6611, pp. 338–349. Springer, Heidelberg (2011). https://doi.org/10.1007/978-3-642-20161-5_34
15. Yin, J., Wang, J.: A Dirichlet multinomial mixture model-based approach for short text clustering. In: SIGKDD (2014)
16. Blei, D.M., Lafferty, J.D.: Correlated topic models. In: NIPS (2005)
17. Yan, X., Guo, J., Lan, Y., Chen, X.: A biterm topic model for short texts. In: WWW (2013)
18. Wang, C., Blei, D.M.: Collaborative topic modeling for recommending scientific articles. In: SIGKDD (2011)
19. Hong, L., Yin, D., Guo, J., Davison, B.D.: Tracking trends: incorporating term volume into temporal topic models. In: SIGKDD (2015)
20. Mikolov, T., Yih, W., Zweig, G.: Linguistic regularities in continuous space word representations. In: Proceedings of the 2013 Conference of the North American Chapter of the Association for Computational Linguistics: Human Language Technologies (2009)
21. Agirre, E., Alfonseca, E., Hall, K., Kravalova, J., Paşca, M., Soroa, A.: A study on similarity and relatedness using distributional and wordnet-based approaches. In: Proceedings of NAACL (2009)
22. Harris, Z.: Distributional structure. Word **10**(23), 146–162 (1994)
23. Liu, Y., Liu, Z., Chua, T.-S., Sun, M.: Topical word embeddings. In: Twenty-Ninth AAAI Conference on Artificial Intelligence (2015)
24. Reisinger, J., Mooney, R.J.: Multi-prototype vector-space models of word meaning. In: Proceedings of HLT-NAACL (2010)
25. Newman, D., Karimi, S., Cavedon, L.: External evaluation of topic models. In: Proceedings of ADCS, pp. 11–18 (2009)

BSBM+: Extending BSBM to Evaluate Annotated RDF Features on Graph Databases

Le Zhang[1], Tong Ruan[1], Haofen Wang[2,3]($\boxtimes$), Yuhang Xia[1], Qi Wang[1], and Dong Xu[4]

[1] East China University of Science and Technology, Shanghai 200237, China
sfrozha@gmail.com, ruantong@ecust.edu.cn,
153996626@qq.com, dsx4602@163.com
[2] Shenzhen Gowild Robotics Co. Ltd., Shenzhen, China
wang_haofen@gowild.cn
[3] Tianjin Key Laboratory of Cognitive Computing and Application, Tianjin, China
[4] North China Sea Marine Technical Support Center, SOA, Qingdao 266033, China
jl6461@aliyun.com

Abstract. Nowadays, more and more knowledge are published in form of RDF triples enriched with numerous types of annotations such as provenance, temporal and geospatial information. Due to the popularity of the ever-growing annotations, graph databases have proposed various storage engine to store these data and extended their query engine to support queries with these annotation constraints. The developers may be curious about the performance of different engines. Regarding the lack of such a benchmark, we develop the first benchmark for this purpose by extending BSBM (one of the most widely used graph database benchmark). We formalize the annotated RDF into a data model with well-defined categories of annotations and their corresponding operators to be supported. Then we extend the data set of BSBM to allow some triples to be annotated with one or more annotations. We further extend the query set to include annotation constraints in a given query, which can be seen as an extension of SPARQL query. We finally select several popular graph databases for benchmark. The experiment results show for each engine, it performs similarly when queried with different type of annotation constraints. No general database designs a special storage or query plan for a specific type of annotations.

Keywords: Linked data · Annotated RDF · Graph database Benchmark

1 Introduction

The Resource Description Framework (RDF) [10] has been proposed by W3C as the standard format for encoding machine-readable information in the Semantic Web [2]. Supplementary to RDF, the W3C has recommended the declarative SPARQL query language, which can be used to extract information from RDF

© Springer Nature Singapore Pte Ltd. 2017
J. Li et al. (Eds.): CCKS 2017, CCIS 784, pp. 129–143, 2017.
https://doi.org/10.1007/978-981-10-7359-5_13

graph bases upon a graph matching facility. In order to enhance the description ability of RDF, [19] firstly proposed Annotated RDF (aRDF) in which RDF triples are annotated by members of a partially ordered set (with bottom element). Based on [19], many researchers develop their own extensions. Meanwhile, SPARQL has been extended to include new operators to fulfill the user information needs on querying these annotations together with the data. The aRDF in this paper is formally defined in Sect. 3.2.

On the other hand, a large number of knowledge bases (KBs) have been published, like DBpedia[1] and Freebase.[2] Various graph engines (GEs) also have been developed to manage these KBs. As introducing various types of annotations to RDF has become popular, some KBs have been enriched with a growing amount of knowledge with meta information. For example, YAGO3[3] (a well-known KB) has annotated its data with meta facts, such as provenance, creator and valid time. Some works [6,20] have extended graph engines to support aRDF and extended SPARQL (eSPARQL) by implementing new operators and special index.[4]

Several efficient evaluations of SPARQL have been proposed, including language-specific framework [5,16], use-case-specific framework [3], and scenarios based on real-world dataset [14]. Unfortunately, we claim that none of them are adequate for testing the annotation features compared with the matured general graph database (gGDB) benchmarks. Thus, to measure these performances and give reliable suggestions in choosing eSPARQL and eGEs, a benchmark for the annotation features of SPARQL is desired. To expand the annotation features evaluation and with respect to the wide use and extensibility of BSBM, we propose the first language-specific benchmark BSBM+ to evaluate annotation features on GEs by extending BSBM to annotation field.

In contrast to general benchmarks, BSBM+ focuses on the annotation features of SPARQL, extensions of SPARQL and extensions of GEs. Thus, we mainly target two goals, GE extensions evaluation and SPARQL extensions evaluation. Specifically, we try to answer the following questions:

1. How different is the support of operators handling annotations on different GEs?
2. How different is the performance of GEs while dealing with different complexities of eSPARQL or different sizes of aRDF?
3. Is it different between the cost of annotation constraints (aCs) and general constraints (gCs)?
4. How does the performance change with the increase of aCs in eSPARQL?

---

[1] http://wiki.dbpedia.org/.

[2] https://www.freebase.com/.

[3] www.mpi-inf.mpg.de/departments/databases-and-information-systems/research/ yago-naga/yago/.

[4] We call SPARQL and graph engines which not support aRDF as general SPARQL (gSPARQL) and general graph engines (gGEs) respectively. In the contrast, we call those which support aRDF as extended SPARQL (eSPARQL) and extended graph engines (eGEs). Especially, we summarize these extensions as annotation features.

5. How does the performance change with the increase of gCs in eSPARQL?
6. How different is the performance affected by different equivalent representations of aRDF?
7. Which selected GE has the best support for aRDF?

In BSBM+, we formalize the aRDF into a well-defined data model and further extend BSBM in two aspects, including datasets and experiments. We first define a new aRDF data model and classify those various annotations into three categories, namely Time, Location and Others. Then we extend the BSBM original dataset by self-annotating and convert aRDF data into different representations. Based on the three categories of annotations, we further design two modules of experiments according to our goals. We rewrite queries in the corresponding forms.

In all, we introduce our work with the following key contributions:

1. We formalize the aRDF into a data model with well-defined categories of annotations and their corresponding operators to be supported.
2. We present the first benchmark on annotation features of SPARQL by annotating the BSBM original dataset and designing two modules of experiments. On this basis, we also convert aRDF data and extended queries into equivalent representations in order to suit different GEs, which do not support aRDF.
3. We apply our BSBM+ to selected GEs and discuss their strengths and weakness associated to annotations that follow from the benchmark results.

The rest of this paper is organized as follows. Section 2 introduces our data model. Section 3 explains the workflow of benchmark. Section 4 shows our dataset extended in detail, while Sect. 5 shows the extended experiments. The results are shown in Sect. 6. Section 7 lists the related work, and Sect. 8 concludes the paper.

2   Related Work

Our work is mainly related to **aRDF**, which is our target data model to benchmark on, and **benchmarks**, which are our basic work. As we focus on special operators for Time and Location, our work is also related to **temporal and geospatial operators**.

2.1   Annotated RDF

Andrea et al. proposed tGRIN indexing structure to index for temporal annotation in [15]. Tappolet et al. designed an extension of SPARQL called τ-SPARQL to query temporal information in [18] and proposed an algorithm to transform τ-SPARQL to standard SPARQL. Recently, Bereta et al. discussed how to represent and query valid time for linked geospatial data in [1]. Mazzieri et al. discussed the semantic of fuzzy RDF in [13]. Straccia et al. discussed the fuzzy extension of ρ df and its joint query in [17]. We proposed fuzzy pD* to further extend the expression ability of fuzzy RDF data in [11] and developed the first reasoning engine to large scale fuzzy data based on MapReduce described in [12]. Olaf Hartig proposed a query language tSPARQL to query trust by introducing two new operators in [7].

However, all the work above only considers one special annotation on RDF data. According to the special characters of the annotation, these work can speed query by rewriting queries and implementing new operators. Thus, these special solutions make these work unable to query RDF data with other types of annotations.

2.2   Benchmarks

Four well-known SPARQL querying benchmarks have been proposed namely the Lehigh University Benchmark (LUBM) [5], the Berlin SPARQL Benchmark (BSBM) [3], the SP^2 Bench SPARQL Performance Benchmark [16] and the DBPedia SPARQL Benchmark (DBPSB) [14]. The former three benchmarks are based on a synthetic dataset and define a scenarios to simulate real world use cases. These benchmarks mainly focus on evaluating the SPARQL performance of tested engines covering all of SPARQL characteristics. The last benchmark, DBPSB, is based on real data from DBPedia. They propose a benchmark creation methodology based on real-world data and query logs. The proposed methodology is used to create a benchmark based on DBPedia data and query logs. All these benchmarks are focus on gSPARQL. In geospatial SPARQL benchmark, Geographica [4] has evaluated geospatial SPARQL on several GEs. These engines support a rich subset of GeoSPARQL and stSPARQL which are two popular geospatial extensions of SPARQL. As for benchmark on data representation, [9] compares the indexing and querying difference of four data reifications.

2.3   Temporal and Geospatial Operators

As for temporal operators, Strabon[5] has implemented strdf to support new temporal datatypes, temporal selections, temporal joins and a rich set of other temporal functions. AllegroGraph[6] also offers some temporal functions (e.g., 'point-beforeInterval'). As for geospatial operators, OGC[7] standards have proposed an extension of SPARQL named GeoSPARQL. WKT and GML are presented to enrich the geospatial information representation. AllegroGraph, Strabon and some other stores have implemented these standards. Except for this extension, Strabon's strdf (an extension of RDF) gives some new geospatial functions. Jena[8] and BlazeGraph[9] also implement some geospatial functions like 'incircle'.

---

[5] www.strabon.di.uoa.gr.
[6] http://franz.com/agraph/allegrograph/.
[7] www.opengeospatial.org/.
[8] https://jena.apache.org/.
[9] https://wiki.blazegraph.com/.

3   Data Model

3.1   Annotation

Various types of annotations have come up, varying from provenance, occurrence-time, to geospatial information. Considering their different storage modes, index demands and supported special operators, it is necessary to formalize them to a united data model with their corresponding operators. We notice that some of these annotations may support different specific operators. We conduct statistics on those operators. For a further usage of our data model, we consider similar usage in the RDF data, as these usages are more likely to be applied into aRDF.

Several works have defined new operators to manipulate **Time**. As mentioned in Sect. 2, τ-SPARQL is designed to support several functions, including 'before', 'after', 'overlap' and others. In the general usage, strdf supports new temporal datatype, like 'period'. stSPARQL enriches temporal selections and temporal joins with a sufficient set of temporal functions. Those works all declare the popularity of temporal information and its query clause's designing trend.

Strabon, uSeek and Parliament have implemented OGC standard to **Location**, which can support various operators like 'east' and 'west'. Strabon's strdf gives some more datatypes on geospatial area, while stSPARQL gives more functions to handle geospatial querying. Jena implements 'incircle' function. Considering the popularity of geospatial information and its complex but helpful usage, geospatial information can be an important part of annotations.

Others of annotations have few widely-used special operators, in both aRDF and RDF. The only work we can find is tRDF [7] include trust into RDF. The 'Using' clause, which is used to retrieve provenance, can also be seen as a kind of provenance annotation usage. Even for the two cases above, they are unitary usages and easy to rewrite, while the usage of Time and Location can be binary. This difference of operations is what we define aRDF according to.

3.2   Annotated RDF

According to the consideration above, Annotated RDF is formally defined as follows.

Definition 2.1

(1) <RDF> ::= <subject>, < predicate>, <object>
(2) <Annotated RDF> ::= <RDF>, <context>
(3) <context> ::= <annotation>{, <annotation>}
(4) <annotation> ::= <predicateAnnotation> : <C>
(5) <C> ::= <Time> | <Location> | <Others>

An RDF triple consists of a subject, a predicate and an object in order. The aRDF in this paper is defined as an RDF triple followed by a context. The context is formed by one or more <annotation>. The <annotation> is a key-value pair, where the key is the predicate of <C> and the value is <C>. The <C> is one of <Time>, <Location> and <Others>. The <Time>, <Location> and <Others> stand for temporal, geospatial and other information, respectively.

For example, <http://dbpedia.org/page/United_States>, <http://dbpedia.org/ontology/leader>, <http://dbpedia.org/page/Bill_Clinton>, {starttime: "1993-01-20"^^xsd:date, endtime: "2001-01-20"^^xsd:date} denotes that Bill Clinton was the leader of United States from January 20, 1993, to January 20, 2001.

For the gGEs, there are four equivalent presentations of aRDF: Standard Reification (SR), N-ary Relations (NR), Singleton Properties (SP) and Named Graph (NG) [8]. Due to space limitations, we do not explain in detail here.

3.3   Extended SPARQL and Extended Graph Engines

In order to query the aRDF data, some extensions extend the gSPARQL syntax and introduce some new special operators corresponding to specific annotations. We call these extensions as eSPARQL. For example, [7] specifies tSPARQL, which extends SPARQL to query the trustworthiness of RDF data. tSPARQL enables users to describe trust requirements in a declarative manner and to access the trust values associated to query solutions. tSPARQL adds two new operators to SPARQL, namely 'TRUST AS' and 'ENSURE TRUST'. The former operator allows access to the trust value associated to the RDF triples, while the latter one allows the range of the trust value to be limited.

Along with eSPARQL, several GEs [6,20] are developed to implement the eSPARQL operators. We call these GEs as eGEs. eGEs are supposed to meet the requirements querying with one or several kinds of annotations. For the example mentioned above, they design tRDF to support tSPARQL by implementing 'TRUST AS' and 'ENSURE TRUST' clauses.

4   Benchmark Workflow

As shown in Fig. 1, BSBM introduces an e-commence dataset with a sufficient schema, and a set of queries focus on gSPARQL use cases. We devise our benchmark by extending BSBM in two aspects: dataset (generated by annotator) and experiments (represented by two query sets).

We extend the dataset based on BSBM original dataset by self-annotating. We annotate Time, Location and Context annotations to the BSBM dataset. Considering that not all the RDF triples have annotations, we introduce a real-world data analyzer to adopt the annotation distribution of YAGO2s to the extended datasets. We then generate three datasets using random distribution, normal distribution and the distribution based on YAGO2s, respectively, as the distribution of annotation values. The datasets' details are shown in Sect. 5. Considering the underlying storage of different engines, we also specially transform our RDF dataset representations.

Then we further design two modules of query set, Engine Extension Module (EEM) and SPARQL Extension Module (SEM). In the EEM, we aim to evaluate

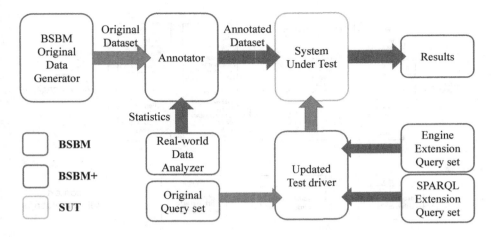

Fig. 1. Workflow of BSBM+ evaluation

the performance between different engines using the same queries. The performance is discussed by partitions respectively focusing on different types of annotations. Each partition is evaluated by a group of queries containing the same type of aCs. Thus, queries of this module are classified into four partitions: Time, Location and Others for single-annotation queries and Mix for multi-annotation queries. In the SEM, we aim to evaluate several aspects of the performance of SPARQL extensions. This module can be divided into two parts. In the first part, queries are organized into three groups with various complexity of aCs and gCs targeting three evaluations. In the second part, we compare the performance of different representations, including native representation and rewritten representations. The experiments' details are shown in Sect. 6. We include a query rewriter to ensure that all these queries can be rewritten into four representations besides the native one, serving four kinds of reification data correspondingly.

In this paper, we also present the result of our first version experiments. For more experiment details, you can refer to our technical report.[10]

5   Dataset

In this section, we present how the extended datasets are generated, as shown in Fig. 2. We use the BSBM generator to generate the raw data. The raw data is a set of RDF triples containing products, producers, product features, product types, vendors, offers, reviewers and reviews. Each RDF triple can be annotated with one or more annotations. The annotations contain Time, Location and Others. The Time consists of starttime and endtime. The Location consists of latitude and longitude. And the Others consists of origin, creator and confidence.

---

[10] www.ecustnlplab.com/papers/BSBM+.pdf.

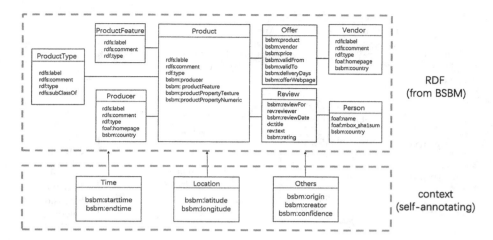

Fig. 2. Schema of annotated RDF datasets

5.1 Distribution of Annotation Type and Annotation Number

Every triple may possibly lose some information. For example, one triple may have creator and confidence, but nothing about when it was created. So, to make our data more reasonable, we adopt the annotation distribution of YAGO2s to datasets, as YAGO2s offers a rich real-world dataset containing various annotations in its MetaFact. We introduce a real-world data analyzer to analyze its annotation distribution, which consists of Annotation Type Ratio ($Ratio_T$) and Annotation Number Ratio ($Ratio_A$). The $Ratio_T$ represents the percentage of each annotation type in the dataset, and the $Ratio_A$ represents the percentage of the triples that have one, two, three, four, five and more annotations. The definitions are given below, and the analysis results are shown in the Fig. 3.

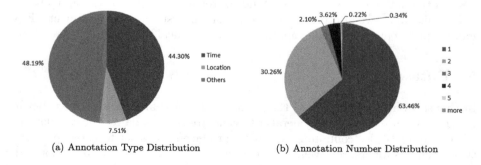

(a) Annotation Type Distribution   (b) Annotation Number Distribution

Fig. 3. Annotation distribution

Definition 4.1. Given an Annotation Type $\in$ {Time, Location, Others}, the Annotation Type Ratio is formally defined as: $Ratio_T(AnnotationType) =$

$\frac{N_T(AnnotationType)}{N}$. Where N_T is the total number of Annotation Type annotations in the dataset. N is the total number of annotations.

Definition 4.2. Given an Annotation Number $\in \{1, 2, 3, 4, 5, more\}$, the Annotation Number Ratio is formally defined as: $Ratio_A(AnnotationNumber) = \frac{N_A(AnnotationNumber)}{N}$. Where N_A is the number of triples that have Annotation Number annotations. N is the total number of annotations.

As the $Ratio_T$ statistics, among 100 triples, there will be 44 Time annotations, 8 Location annotations and 46 Others annotations. In order to meet the multiple annotations requirement, we distribute every kind of annotations as above. The $Ratio_A$ is adopted in the same way.

5.2 Distribution of the Annotation Values

As different distributions of the annotation values may lead to different evaluation results, we prepare different distributions of annotation values for evaluation. We generate three datasets using uniform distribution, normal distribution and the distribution based on YAGO2s, respectively, as the distribution of annotation values. As for the distribution based on YAGO2s, we extract all the annotations from YAGO2s metadata to build an Annotation Candidate Dataset. Then we randomly choose an annotation from the dataset to annotate a triple. Due to space limitations, we do not explain in detail here.

6 Experiment

In this section, we introduce our experiment settings and circumstances of Engine Extension Module and SPARQL Extension Module. The specific queries are not listed here.

6.1 Metrics

With respect to BSBM, we further use some of their metrics in BSBM+ experiments. We use **Queries per Second** (QpS) as an basic metric that measures the number of queries that are answered by GEs within a second. We repeat our execution for a certain time and retrieve the mean of QpSs. Its reciprocal Query Response Time (QRT) is also included as an equivalent metric. We use **Load Time** (LT) to evaluate the ability to load annotated data of graph databases (GDBs). It measures the total time of loading and indexing for a special dataset. LT should always be reported with the size and representation of dataset.

6.2 Graph Engines Chosen and Experiment Circumstance

We choose Jena, Sesame, AllegroGraph, BlazeGraph and tRDF to be tested. We are prepared to evaluate both extended graph databases (eGDBs) and gGDBs. Unfortunately, most research lingers around the theoretical level without implementations. We can seldom find a well-implemented system or prototype except tRDF. So we choose four popular gGDBs to be tested by data representation transforming and query rewriting, in addition to tRDF. In these gGDBs, Jena is version 2.12.1 without NG4J. Sesame is version 2.7.13. AllegroGraph is version 5.1.1, Free Edition. BlazeGraph is version 1.2.4. tRDF version is tRDF4Jena 0.1.3 with Jena ARQ 2.10.1. All experiments are conducted under Linux 2.6.x 64 bit, on top of a server with four Inter(R) Xeon(R) CPU E5-2407 2.20 GHz processors and 16 GB memory. The Java engine is JDK 1.7.

6.3 Engine Extension Module

In this module, we focus on questions 1 to 3 we proposed in the introduction. We compare all the engines in the same circumstance with an incremental size of dataset loaded. We conduct this evaluation to study the performance of loading and query response on different types of annotations among different GDBs. When loaded with datasets of the same size, we run EEM on those systems under test (SUTs). Then we analyze the performances of different annotation categories to show these engines' support for those categories of annotations. When SUTs are queried by queries containing special operators, we can answer question 1. When SUTs are queried by different complexities of queries, we can answer question 2. When the size of the loaded dataset changes, we can answer question 3. Since tRDF has special index and operators, we analyze it separately. QpS (or QRT) and LT are used to evaluate the performance here. Uniform distribution is taken as annotation value distribution.

6.4 SPARQL Extension Module

We use this module of experiments to measure annotation features of eSPARQL, aiming at questions 4 to 7. QpS (or QRT) is used to evaluate the performance here.

In this module, two aspects of queries are included. In one aspect, we compare each engine in the same circumstance with different queries in three groups to answer questions 4, 5 and 6, respectively. We mainly focus on the cost and scalability of annotations disregarding the ability of loading. We use uniform distribution as annotation value distribution in this aspect.

General Constraint Replacements. Given a query with aCs, we replace one aC with a similarly functional gC to generate a new query to determine the efficiency of annotation compared to the gSPARQL. We conduct this group of evaluation to check whether it is more efficient using annotation to tag meta information than using RDF.

Incremental Annotation Constraints. With an original query containing only gCs, we incrementally increase the aC numbers to generate a set of queries with different complexities. By executing this group of evaluation, we can curve the incremental QRT trends as the number of annotation increases.

Incremental General Constraints. With an original query containing gCs and aCs, we incrementally increase the gCs to increase the complexity of gCs and generate a set of queries. By executing this group of evaluation, we can curve the incremental QRT trends as the complexity increases. This group discusses the scalability of aCs.

In other aspect, we compare the difference between four representations as question 7 required. We use uniform, normal and YAGO2s based distributions as annotation value distributions.

Different Representation Comparison. With the same size of dataset, we load four representations into a SUT. After the SUT is queried by a set of specially selected queries, we compare the performance of different representations along with different queries, respectively. We change our target database in experiments as well, since different databases are usually close to different representations.

7   Results

In this section, we present results from Engine Extension Module and SPARQL Extension Module. Due to space limitations, only some interesting figures are described.

7.1   Engine Extension Module

We perform EEM on four gGDBs using uniform distribution. For question 1, as Fig. 4(a) shows, we count the average QpS of four types of queries. Jena performs higher than the other three GDBs, as it has indexed SPO, POS and OSP when data is loaded. Sesame scores second because it uses a native memory store to speed up query performance. Blazegraph and Allegrograph achieve similar performances. Generally speaking, all four types of annotations keep this rank. This shows those gGDBs have little special indexing or underlying storage design for those types of annotations and achieve similar scores on each type.

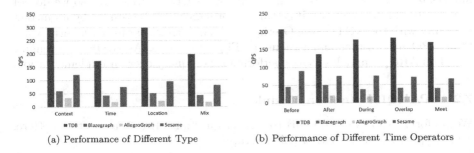

(a) Performance of Different Type   (b) Performance of Different Time Operators

Fig. 4. Engine Extension Module results

As Time type has the lowest score, we specially analyze performance results in Fig. 4(b). We find that all the Time operators receive similar response time, while 'after' and 'before' operators are executed slightly faster. Jena executes 'before' faster than 'after' because it may index its date type in an increasing order.

As for question 2, queries with simple filters are all executed faster than ones with complex filters in the four gGDBs. However, queries with multi-annotations are sometimes executed faster than ones with complex filters. The complexity of filters matters more. Generally speaking, all four GDBs perform in the same rank as the results in question 1, with Jena TDB winning first. To answer question 3, all the GDBs slow down in some degree but keep the same rank.

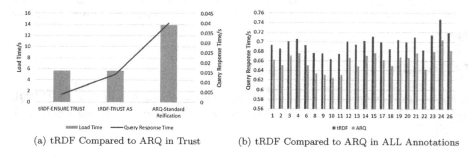

(a) tRDF Compared to ARQ in Trust (b) tRDF Compared to ARQ in ALL Annotations

Fig. 5. Performance of tRDF

Additionally, we compare tRDF with ARQ (a query engine of Jena) in 'Trust' (which we also call 'Confidence') and all other annotations (26 in total), and the results shown in Fig. 5, following question 1. In 'Trust', we rewrite its 'ENSURE TRUST' clause using 'TRUST AS' clause and gSPARQL syntax to generate a similar query. We transform the dataset to SR. Figure 5(a) shows tRDF achieves higher performance both in loading and querying, while its 'ENSURE TRUST' clause outperforms the rewritten 'TRUST AS' clause. When we transform the dataset to NG, it works even worse at 196.415 s LT and 67 s QRT so we do not put the results in the figure. Its awful performance may be caused by its incompatibility with the NG4J suite. On the other hand, tRDF performs slightly worse than ARQ in all other annotations, as Fig. 5(b) shows. As tRDF is an engine based on ARQ, it may lose some performance on query over an implementation.

7.2 SPARQL Extension Module

We perform SEM on each GDB to answer questions 4 to 7, respectively. Here, we only present our results of Allegrograph.

General Constraint Replacements. From Fig. 6(a), when we replace a gC by an aC, the response time grows remarkably, as the number on the x-axis represents complexity of gCs. Using aCs takes a longer than using gCs. When the dataset and original query complexity increasing, the cost disparity also increases.

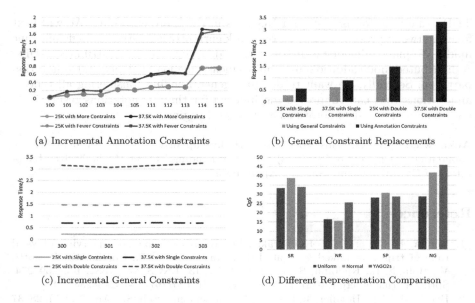

(a) Incremental Annotation Constraints   (b) General Constraint Replacements

(c) Incremental General Constraints   (d) Different Representation Comparison

Fig. 6. SPARQL Extension Module results

Incremental Annotation Constraints. In Fig. 6(b), the numbers on the x-axis represent the id of queries. A larger id means the query contains more aCs. The performance is similar with more and fewer gCs. Compared to gCs, aCs have more effects on query response time. As aCs increase, response time grows rapidly. Meanwhile, as dataset size grows, response time increases significantly.

Incremental General Constraints. In Fig. 6(c), we find the performance remains stable, as the number on the x-axis represents complexity of gCs. Containing more gCs slightly influences the performance, compared to Fig. 6(a). Dataset size and aCs complexity mainly determine the response performance.

Different Representation Comparison. In Fig. 6(d), NG generally gets the best result, as its structure is the most concise and its schema has a longer length. SR performs secondly, and SP follows it. NR performs poorly since its structure is complex and may not be suitable for annotations.

8 Conclusion and Future Work

In this paper, we formalize the aRDF into a data model with three categories of annotations. We propose BSBM+ for evaluating the annotation features on popular GEs. The results show special indexed engines can perform well in a specific annotation type but not in other types. As for gGDBs, they can only execute queries with aCs at a general data level efficiency, or even worse considering the complexity of aRDF representations, because these databases have few special indexes and other underlying storages design for annotated RDF. Jena

TDB usually performs best, and NG wins in the representations. We will next focus on two improvements. First, our work currently is limited to single machine mode. We plan to include a distributed mode to evaluate on distributed GDBs. A distributed mode additionally requires measuring load balance and annotation partition. Secondly, we want to design a general engine considering all type of annotations that can support being indexed for three annotations respectively and to support most annotations as a general engine.

Acknowledgements. This work was partially supported by the National Science Foundation of China (No: 61402173) and Open Funding Project of Tianjin Key Laboratory of Cognitive Computing and Application.

References

1. Bereta, K., Smeros, P., Koubarakis, M.: Representation and querying of valid time of triples in linked geospatial data. In: Cimiano, P., Corcho, O., Presutti, V., Hollink, L., Rudolph, S. (eds.) ESWC 2013. LNCS, vol. 7882, pp. 259–274. Springer, Heidelberg (2013). https://doi.org/10.1007/978-3-642-38288-8_18
2. Berners-Lee, T., Hendler, J., Lassila, O.: The semantic web. Sci. Am. **284**(5), 28–37 (2001)
3. Bizer, C., Schultz, A.: The Berlin SPARQL benchmark (2009)
4. Garbis, G., Kyzirakos, K., Koubarakis, M.: Geographica: a benchmark for geospatial RDF stores (long version). In: Alani, H., Kagal, L., Fokoue, A., Groth, P., Biemann, C., Parreira, J.X., Aroyo, L., Noy, N., Welty, C., Janowicz, K. (eds.) ISWC 2013. LNCS, vol. 8219, pp. 343–359. Springer, Heidelberg (2013). https://doi.org/10.1007/978-3-642-41338-4_22
5. Guo, Y., Pan, Z., Heflin, J.: LUBM: a benchmark for owl knowledge base systems. Web Seman. Sci. Serv. Agents World Wide Web **3**(2), 158–182 (2005)
6. Harth, A., Umbrich, J., Hogan, A., Decker, S.: YARS2: a federated repository for querying graph structured data from the web. In: Aberer, K., Choi, K.-S., Noy, N., Allemang, D., Lee, K.-I., Nixon, L., Golbeck, J., Mika, P., Maynard, D., Mizoguchi, R., Schreiber, G., Cudré-Mauroux, P. (eds.) ASWC/ISWC -2007. LNCS, vol. 4825, pp. 211–224. Springer, Heidelberg (2007). https://doi.org/10.1007/978-3-540-76298-0_16
7. Hartig, O.: Querying trust in RDF data with tSPARQL. In: Aroyo, L., et al. (eds.) ESWC 2009. LNCS, vol. 5554, pp. 5–20. Springer, Heidelberg (2009). https://doi.org/10.1007/978-3-642-02121-3_5
8. Hartig, O., Thompson, B.: Foundations of an alternative approach to reification in RDF. arXiv preprint arXiv:1406.3399 (2014)
9. Hernández, D., Hogan, A., Krötzsch, M.: Reifying RDF: what works well with wikidata? In: ISWC 2015, Bethlehem, PA, USA, pp. 32–47 (2015)
10. Klyne, G., Carroll, J.J., McBride, B.: Resource description framework (RDF): concepts and abstract syntax. World Wide Web Consortium Recommendation (2004)
11. Liu, C., Qi, G., Wang, H., Yu, Y.: Fuzzy reasoning over RDF data using owl vocabulary. In: Proceedings of the 2011 IEEE/WIC/ACM ICWIIAT, pp. 162–169. IEEE Computer Society (2011)
12. Liu, C., Qi, G., Wang, H., Yu, Y.: Reasoning with large scale ontologies in fuzzy pD* using mapreduce. IEEE Comput. Intell. Mag. **7**(2), 54–66 (2012)

13. Mazzieri, M., Dragoni, A.F.: A fuzzy semantics for the resource description framework (2008)
14. Morsey, M., Lehmann, J., Auer, S., Ngonga Ngomo, A.-C.: DBPedia SPARQL benchmark-performance assessment with real queries on real data. In: ISWC 2011, pp. 454–469 (2011)
15. Pugliese, A., Udrea, O., Subrahmanian, V.: Scaling RDF with time. In: Proceedings of the 17th International Conference on World Wide Web, pp. 605–614. ACM (2008)
16. Schmidt, M., Hornung, T., Lausen, G., Pinkel, C.: SP^{2}Bench: a SPARQL performance benchmark. In: Data Engineering, 2009. ICDE 2009, pp. 222–233 (2009)
17. Straccia, U.: A minimal deductive system for general fuzzy RDF. In: Polleres, A., Swift, T. (eds.) RR 2009. LNCS, vol. 5837, pp. 166–181. Springer, Heidelberg (2009). https://doi.org/10.1007/978-3-642-05082-4_12
18. Tappolet, J., Bernstein, A.: Applied temporal RDF: efficient temporal querying of RDF data with SPARQL. In: Aroyo, L., et al. (eds.) ESWC 2009. LNCS, vol. 5554, pp. 308–322. Springer, Heidelberg (2009). https://doi.org/10.1007/978-3-642-02121-3_25
19. Udrea, O., Recupero, D.R., Subrahmanian, V.S.: Annotated RDF. In: Sure, Y., Domingue, J. (eds.) ESWC 2006. LNCS, vol. 4011, pp. 487–501. Springer, Heidelberg (2006). https://doi.org/10.1007/11762256_36
20. Weiss, C., Karras, P., Bernstein, A.: Hexastore: sextuple indexing for semantic web data management. Proc. VLDB Endow. 1(1), 1008–1019 (2008)

Detecting Spammers in Sina Micro-blog Based on Multiple Features

Zhenwei Gao[✉], Mingwei Tang, Yajun Du, and Shimin Zhong

School of Computer and Software Engineering, Xihua University,
Chengdu, Sichuan 610039, China
gao13628026637@gmail.com

Abstract. As the explosive growth of micro-blog users, which resulting in a large number of spammers. Not only for the calculation of user influences and social network relationship analysis has brought new challenges, and spammers led to the emergence of a serious crisis of confidence in microblog development. A number of research works have been carried out to solve these issues but most of the existing techniques had not focused on various features and doesn't group similar user micro-blog contents which become their major limitation. In this paper, we focus on those who send a large number of advertising posts spammer users. We combine users' relationship features, behavior features and tweets' content features together and then proposes advertising spammers detecting model (ASDM), The experimental results show that this system is better than that of other systems.

Keywords: Spammers detecting · ASDM · Social network · Multiple features

1 Introduction

Micro-blog is one of the most popular online social networking (OSN) and micro-blog services that enables its users to send and read contents based posts of up to 140 characters, known as "posts". Nowadays, millions of users use Sina micro-blog to keep in touch with friends, meet new people and discuss about everything. In a social network, people exchange a great deal of information and it is common to observe that certain individuals have especially strong influences on others. The number of followers reflects the user's influence and popularity.

Z. Gao—This work was supported by the National Natural Science Foundation of China under Grant (No. 61472329, 61532009 and 61602389), the key of Scientific Research Funds Project of Educational Department of Sichuan Province (No. 13ZA0019), the Scientific Research Funds Project of Science and Technology Department of Sichuan Province (No. 2016JY0244, 03022044), the Scientific Research Funds Project of "Chun Hui" plan of Ministry of Education of China (No. z2014051). Sichuan outstanding youth science and Technology Fund training program (No. 2017JQ0059).

J. Li et al. (Eds.): CCKS 2017, CCIS 784, pp. 144–151, 2017.
https://doi.org/10.1007/978-981-10-7359-5_14

In order to quickly increase the number of fans to meet vanity, a large number of fans buying and selling phenomenon on the network. Legitimate users can not meet the demand, so the machine registered spammers began to appear, it was used to "brush", "sale" and so on, to sell the goods by false advertising infringement, normal the interests of users, micro-blog will also reduce the credibility and dirty network environment, therefore, the recognition and clearance of spammers has the important practical significance for the benign development of the micro-blog community.

At present, there are a lot of researches on micro-blog in academia, but there are few literatures on the research of spammers, existing methods mainly use the characteristics of manual search accounts, such as the account name, the number of fans, concerned about the number of features, and analysis features of sample data, carries on the training corpus using machine learning algorithm to identify the data of spammers, because Sina official crackdown on spammers and more spammers appear more intelligent than before, the new type of spammers, the user presents some new characteristics, on the basis of the previous simple rules of discrimination ability is more and more weakly.

Spammers users have many differences with legitimate users, for instance, content features of posts, users' behavior features and relationship features. In previous studies, a majority of researchers adopted only one kind of these features. For example, Fu et al. [1] proposed the problem of spammer detection by leveraging the "carefulness" of users. They propose a framework to measure the carefulness, and develop a supervised learning algorithm to estimate it based on known spammers and legitimate users. Lim et al. [2] identify several characteristic behaviors of review spammers and model these behaviors so as to detect the spammers. Unfortunately, a single feature can not effectively distinguish between spammers and legitimate users. Micro-blog as a new social platform, in recent years a large number of advertising spammers for the emergence of micro-blog credibility of the great challenge. However, researchers hardly have studied advertising spammers in Micro-blog. In this paper, we focus on those who send a large number of advertising posts spammer users. we combine users' relationship features, behavior features and tweets' content features together and then proposes advertising spammers detecting model (ASDM), the experimental results show that this system is better than that of other systems.

The rest of the paper is organized as the following: In Sect. 2, we give an overview of related works. In Sect. 3, we will introduce our method in detail. In Sect. 4, we will use the data set to elaborate on our experiments. In Sect. 5, we conclude this work and present future work.

2 Related Work

The success of Sina Micro-blog has attracted the attention of many researchers. Previous detection spammers was based on some simple rules. For example, Ming and Yi [3] proposes a new intelligent classification method based on the characteristics of registered usernames. The social relations of spammers in social network are different from legitimate users. For example, the number of spammers' followees is much larger than their friends. Xiying et al. [4] proposes using the machine learning algorithm combined with the features selected from the user profile and tweets, such as the number of followers, the number of posts. Gao et al. [5] proposes a variety of features to distinguish between advertising spammers, following spammers and legitimate users. Han et al. [6] proposes a probabilistic graphical model named spammers probabilistic graphical model (SPGM) in which user profile features are treated as the input variables while behaviour features act as the observed variables and the probability of a user being a spammer is a hidden variable of SPGM model. Spammers are also different from legitimate users in behavior features. For example, they retweet other users' posts more often than posts by themselves. As a result, some experts put forward methods based on users' behavior features to detect spammers. Yanmei et al. [7] proposes a classifier based on the behavior characteristics using Bayesian model and genetic algorithm to distinguish the spammers. Wang et al. [8] proposes a suitability of five classification algorithms and four different feature sets to the social spam detection task. Soman and Murugappan [9] detecting malicious tweets in trending topics using clustering and classification. Jindal and Liu [10] using repetitive detection techniques to identify spam reviews in the review system, Draisbach et al. [11] using the adaptive time window to detect the repeat ability of the text list. Zhao et al. [12] using reuse detection to identify micro-blog spam messages. Zhang et al. [13] based on the Hadoop platform and repeat the detection algorithm, screening the social network platform for spam messages.

Previous methods mainly consider the characteristics of the user's attention to the number, the number of fans, forwarding and account to identify spammers, however, spammers become more and more intelligent, the previous method to identify the new spammers there is a lot of misjudgment phenomenon. Micro-blog as a new social platform, in recent years a large number of advertising spammers for the emergence of micro-blog credibility of the great challenge. However, researchers hardly have studied advertising spammers in Micro-blog. In this paper, we focus on those who send a large number of advertising posts spammer users. We combine users' relationship features, behavior features and tweets' content features together and then proposes advertising spammers detecting model (ASDM), the experimental results show that this system is better than that of other systems.

3   Advertising Spammers Detecting Model

As the advertising spammers will be in a period of time to send a lot of content to repeat or similar Micro-blog posts. We divided the tweets into the original tweets' and retweets, each of which is retweets to form a tweets' list, and apply repetitive detecting to this tweets' list. Advertising spammers detecting model can be described as follow:

(1) If the user u retweets the user v issued, the user v constitutes the retweets source user set V, for each $u, v \in V$. According to the time sequence of the user to create the original micro-blog L_{vv} as well as u retweets v micro-blog sequence L_{uv}.

(2) We prepare each micro-blog sequence in micro-blog sequence, remove some useless information, and then use the LDA theme model to get the topic vector for each post. Finally, calculate the similarity of two consecutive posts.

(3) For the previous L_{vv}, L_{uv}, we use our definition of the degree of repeat evaluation function to calculate the repeat of the consecutive posts.

(4) The sum of the repetitions of each pair of posts in L_{uv} and combined with L_{vv}. Finally, the number of retweets is taken into account. We then proceed to normalize and identify advertising spammers.

3.1 Basic Definitions

The advertising spammers detecting the function:

$$F_u(u) = \frac{\sum F_{uv}(v)}{|V| + 1} + F_{uu}(u) - SC(u) * \Phi(t) \tag{1}$$

$F_{uv}(v)$ is the repeat evaluation function of the micro-blog sequence of u retweets v. V represent the user v constitutes the retweets source user set, $F_{uu}(u)$ represents the repetitive evaluation function of the original micro-blog published by the user u. $SC(u)$ represents posts count of u. $\Phi(t)$ is the probability of spammers tweets for each subject dimension from the training set data. ASDM mainly consider the user's retweets and tweets repeatability and posts contents information.

$F_{uv}(v)$ is the repeat evaluation function:

$$F_{uv}(v) = F_{uu}(u) \times F_v(v) \tag{2}$$

The micro-blog sequence repeatability evaluation function $F_{uu}(u)$ can be defined as follows:

$$F_{uu}(u) = \frac{\sum_{i=1}^{N_m-1} F_m(A, B)}{N_m - 1} \tag{3}$$

N_m is the number of posts. $F_m(A, B)$ is a function of evaluating the repeatability of two consecutive posts, it will be described in detail in the next section, next we define the user attribute evaluation function $F_v(v)$, it can be defined as follows:

$$F_v(v) = \left(1 + \frac{FC(v)}{\sum_{v \in V} FC(v)} + \frac{SC(v)}{\sum_{v \in V} SC(v)}\right) \tag{4}$$

$FC(v)$ is the number of followee of the user v, $SC(v)$ is the number of tweets counts, V represent the user v constitutes the retweets source user set, $F_v(v)$ is the function that defines the importance of the source user, the greater the $F_v(v)$, the greater the impact on the model.

3.2 Similarity Detection

As Sina micro-blog contents is short, less useful information, micro-blog topic change-able, hot events a lot, may cause a higher dimensionality. For contents short features, there are two general approaches to academia and industry: the first is to extend short contents with external contents such as search engine results; the second is to use the knowledge base, such as WordNet or Wikipedia, to dig short contents the inner rela-tionship of words. The machine learning algorithm used for text classification, for SVM, Bayes, KNN, LLSF and decision tree and so on. In this paper, the LDA model is used to model the micro-blog short text from the thematic dimension. With the LDA theme model, we can get the topic distribution for each post, t_u^a represents the distribution of the topic set of the post A, A and B $\left(t_u^b\right)$ are two consecutive posts. $A, B \in N_m$, N_m is a set of user posts. The similarity of two topic collections A and B are measured by Jensen–Shannon divergence, which is defined as:

$$\text{TopicSimilarity} = S_{AB} = \frac{1}{2} \times D_{KL}\left(t_u^a || M\right) + \frac{1}{2} \times D_{KL}\left(t_u^b || M\right) \tag{5}$$

where D_{KL} is the Kullback–Leibler divergence, which is defined as $D_{KL}\left(t_u^a || t_u^b\right) = \sum \ln\left(\frac{t_u^a}{t_u^b}\right) \times t_u^a$, and $M = \frac{1}{2}\left(t_u^a + t_u^b\right)$ Jensen–Shannon divergence which is a popular method of measuring the similarity between two probability distributions.

As the spammer users will send a lot of advertising posts in a short time, so the posting interval is also the factors. We can use t_1, t_2 to express the posts A and B posting time. $F_m(A, B)$ can be defined as follows:

$$F_m(A, B) = \frac{S_{AB}}{\ln|t_2 - t_1| + 1} \times F_s(A, B) \tag{6}$$

$F_m(A, B)$ is a function of evaluating the repeatability of two consecutive posts. $F_s(A, B)$ is the advertising text probability assessment function, it assessed the informa-tion in the text, the statistics of the text in the probability of advertising posts. It can be defined as follows:

$$F_s(A, B) = \psi(A, B) \times \Phi(t) \tag{7}$$

$\psi(A, B)$ is the topic vector in the micro-blog text. $\Phi(t)$ is the probability of spammers tweets for each subject dimension from the training set data. Which is can be defined as follows:

$$\Phi(t) = \frac{\sum\limits_{adspammer} s}{\sum\limits_{all} T} \tag{8}$$

$\Phi(t)$ is the probability of advertising posts for each subject dimension from the training set data. s is the topic of advertising spammers distribution, T is the topic distribution in the training data set.

4 Experiment and Analysis on Dataset

4.1 Dataset and Preprocessing

A dataset crawled from Sina Weibo via Sina Weibo API is used in this study. We crawl the 782 users a total of 48814 posts. First clean the dataset, we get 48513 effective posts. Finally, we put the dataset of posts into retweet posts and original posts, the statistics are shown in Table 1.

Table 1. Statistics of the retweet posts and the original posts

Type	Number of posts	Source users
Retweet posts	18701	3814
Original posts	29812	782

We divided the data set into two parts, 60% as the training set, and 40% as the test set. We use the artificial standard way to mark the training data set of advertising posts. Then use the LDA theme model for the advertising posts and random user posts. We have counted the topics distribution of the highest probability of random users and advertising users, the statistics are shown in Table 2.

Table 2. Statistics of random users and advertising users topics distribution

Ranking	Random users	Advertising users
1	Haizi, mama, baba, baobao, nanren, erzi shuaige, meinv	Choujiang, weixin, daili, dujia, mianfei, pifa, Vxin, lianjie
2	Shenghuo, cuozhe, rensheng, women, shiqing, ziji	Hunsha, sheying, xiqing, xinren, dapei, jingmei
3	Shijie, lvyou, yujian, yiqi, fengjing, meili. mingxing	Dingzhi, shipai, daifa, xianliang, Shili, baokuan
4	Jingli, yinhang, gupiao, touzhi, shichang, licai, gongsi	Tuijian, dizhi, dianji, xiangqing, Xinkuan, paoliang
5	Xinwen, shehui, guojia, renmin Shijian, lvshi	Fuzuoyong, meibai, xiaoguo, Duanhuo, quban, meirong

From Table 2 can be seen, advertising users and random users topics distribution has a big difference, that means our method is effective.

4.2 Evaluation Criterion

We evaluate the advertising spammer detection performance on our dataset by using ASDM and two machine learning algorithms, Naive Bayes [7] and Support Vector

Machine [4]. We select the microblogging users concerned about the ratio of their fans, microblogging users all with the url link ratio, mutual concern, average number of comments as the feature of SVM and Naive Bayes training. For machine learning algorithms we adopted 10-fold cross-validation to reduce the generalization error. In order to evaluate the experimental result, we selected the following evaluation criterions Precision, Recall and F-value.

In Fig. 1, we can see that, the Advertising Spammers Detecting Model is the best at precision, the precision rate, recall rate and F-measure rate of all the classifiers reached more than 91%. We mainly center on the value of F-value to evaluate the classifiers as it is a comprehensive metric of summarizing both precision and recall. From Fig. 1, we can see the Advertising Spammers Detecting Model has the best capability among the three tested classifiers. Due to the sparseness of the text and the presence of some noise and the emergence of microblogging network words, the recall rate of ASDM is lower than SVM, but it is better than naive Bayes.

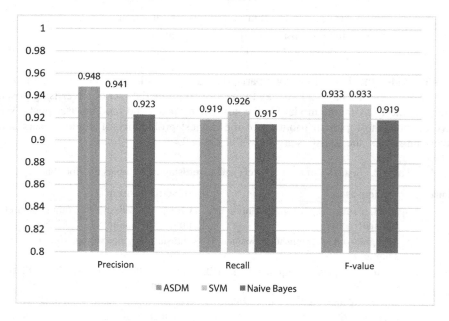

Fig. 1. Comparison of different methods

5   Conclusions and Future Work

In this paper, we focus on those who send a large number of advertising posts spammer users. We combine users' relationship features, behavior features and tweets' content features together and then proposes advertising spammers detecting model. We found that advertising spammers have many different features, for example, advertising spammers post many tweets in order to spread advertisements per day, but advertising spammers rarely retweet other users posts. Based on these differences, we use our model to detect advertising spammers, to evaluate the spammer detection performance on our dataset, we use the machine learning algorithm to compare with our model. The experimental results show that this system is better than that of other systems.

There are some limitations of this study which deserve intensive future research. In this paper, we focus on those who send a large number of advertising posts spammer users. We combine users' relationship features, behavior features and tweets' content features together and then proposes advertising spammers detecting model. As advertising spammer users in order to avoid Sina's detection, advertising words more and more vague expression, so the use of topic distribution to detect the effect of advertising users is not very good. These limitations can promote future research.

References

1. Fu, H., Xie, X., Rui, Y.: Leveraging careful microblog users for spammer detection. In: International Conference on World Wide Web, pp. 419–429. ACM (2015)
2. Lim, E.P., Nguyen, V.A., Jindal, N., et al.: Detecting product review spammers using rating behaviors. In: ACM International Conference on Information and Knowledge Management, pp. 939–948. ACM (2010)
3. Ming, F., Yi, F.: A new intelligent recognition method of zombie fan. Comput. Eng. **39**(4), 190–193 (2013)
4. Xiying, Z., Xin, C., Xianyun, T.: A recognition method of zombie fans on micro-blog user's behavior. J. Nat. Sci. Heilongjiang Univ. **31**(2), 250–254 (2014)
5. Gao, S., Ma, X., Wang, L., et al.: Spammer detection based on comprehensive features in Sina Microblog. In: International Conference on Service Systems and Service Management, pp. 1–6. IEEE (2016)
6. Han, Z., Yang, K., Xu, F., et al.: Probabilistic graphical model for detecting spammers in microblog websites. Int. J. Embed. Syst. **8**(1), 12 (2016)
7. Zhang, Y., Huang, Y., et al.: Weibo spammers' identification algorithm based on Bayesian model. J. Commun. **1**(38) (2017)
8. Wang, B., Zubiaga, A., Liakata, M., et al.: Making the most of tweet inherent features for social spam detection on twitter. In: WWW Workshop on Making Sense of Microposts (2015)
9. Soman, S.J., Murugappan, S.: Detecting malicious tweets in trending topics using clustering and classification. In: International Conference on Recent Trends in Information Technology, pp. 1–6. IEEE (2014)
10. Jindal, N., Liu, B.: Review spam detection. In: International Conference on World Wide Web, pp. 1189–1190 (2007)
11. Draisbach, U., Naumann, F., Szott, S., et al.: Adaptive windows for duplicate detection. In: IEEE International Conference on Data Engineering, pp. 1073–1083. IEEE (2012)
12. Zhao, B., Ji, G.L., Qu, W.G., et al.: Detecting microblog spammers based on reuse detection. J. Nanjing Univ. **49**(4), 456–464 (2013)
13. Zhang, Q., Ma, H., Qian, W., et al.: Duplicate detection for identifying social spam in microblogs. In: IEEE International Congress on Big Data, pp. 141–148. IEEE (2013)
14. Wu, B., Li, G., Liu, Y., Zhang, L., Wang, B.: Spammer detection based on duplicate microblog post. J. Data Acquis. Process. **30**(1), 117–125 (2015)

A Hybrid Method to Sentiment Analysis for Chinese Microblog

Xia Fu[✉], Yajun Du, and Yongtao Ye

School of Computer and Software Engineering, Xihua University,
Chengdu 610039, China
1805256845@qq.com

Abstract. In recent years, more and more netizens are willing to express their opinions on social media platforms. Sentiment analysis is effective and valuable to extract useful information out of massive text documents. In this paper, we proposed a hybrid approach to the sentiment analysis problem for Chinese microblog. This hybrid approach combines the basic techniques of natural language processing (NLP) and machine learning to determine the semantic orientation for Chinese microblog. The hybrid method is tested on two public data sets and the results show that our method is effective.

Keywords: Sentiment analysis · Machine learning · Word embedding

1 Introduction

Sentiment analysis (SA) also called opinion mining, refers to the progress of analyzing, processing and summing up for subjective information with emotional color. At present there are many researchers have done works on the sentiment analysis, the technology is mainly divided into two categories: one is lexicon-based approach; the other is machine learning approach. Lexicon-based method can not solve the SA problem very well, because of the low coverage of emotional dictionary and the lack of emoticons, at the same time, lexicon-based method can not solve the unknown words. Machine learning method also can not solve the SA problem very well, because it rely on manual annotation too much.

In this paper, considering the speciality of Chinese microblog and the drawbacks of traditional methods, we proposed a hybrid method to deal with the sentiment analysis problem. The main contributions of this paper are summarized as follows:

(1) We take advantage of the strengths of lexicon-based, machine learning and word embeddings approach, meanwhile, overcome the weakness of these three methods.
(2) In the process of solving the problem we considering the emoticons.

© Springer Nature Singapore Pte Ltd. 2017
J. Li et al. (Eds.): CCKS 2017, CCIS 784, pp. 152–157, 2017.
https://doi.org/10.1007/978-981-10-7359-5_15

The remainder of the paper is organized as follows: Sect. 2 discussed the related work; Sect. 3 shows the hybrid method proposed in this paper; the experiment results are displayed in Sect. 4; Finally, Sect. 5 talked about the conclusion and further work.

2 Related Work

Sentiment analysis can be divided into words, phrases, sentences and documents level etc. Since Sentiment analysis proposed by Pang et al. [1], there has been a large amount of work on sentiment analysis at various levels of granularity. These researches fall into two categories: lexicon-based and machine leaning method.

2.1 Lexicon-Based

Lexicon-based method mainly divided into the following steps: doing syntactic analysis and part-of-speech tagging, etc. for documents by formulating a series of emotional dictionary and rules; computing emotional values; judging the sentiment origination of the documents according to the sentiment values. Shen et al. [2] calculated emotional value of microblog by building negative, degree and interjection dictionary. Hung and Chen [3] build a new SentiWordNet lexicon. With the human language becoming ambiguous, the exact meanings of a word in SentiWordNet needs to be justified according to the context in which the word occurs.

2.2 Machine Leaning Method

Supervised machine learning (ML) methods can take advantage of annotated corpus and machine learning models. Considering the superiority of machine learning method, Pang et al. [1] treat sentiment analysis as a binary classification problem. Tripathy et al. [4] use various supervised machine learning, such as NB, ME, Stochastic Gradient and SVM to classify move reviews. The result shows that the value of n in n-gram increases the classification accuracy. Katz et al. [5] proposed a novel context-based method. They models the sentiment terms and the contexts in which they appear to generate features for supervised learning. The results show that their method has strong robustness.

Different from supervised method, unsupervised method does not use labeled documents, but depend on document's statistical properties (e.g. word occurrence). Turney [6] apply the unsupervised ML method to sentiment analysis. They proposed a PMI (point mutual information)-based method to analyze the emotional polarity of specific phrase. The unsupervised method is simple, but it depends on the domain of processing corpus and particularly rely on basic words.

3   A Hybrid Method to the Sentiment Analysis for Chinese Microblog (SAFCM)

The graphic depiction of the framework of the hybrid method is shown in Fig. 1.

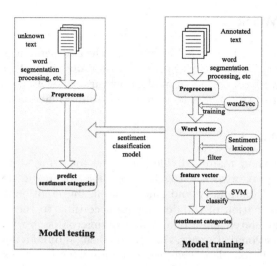

Fig. 1. The framework of the hybrid method.

3.1   Construction of Sentiment Lexicon

The sentiment orientation of text main reflected by sentiment words. The coverage ratio of sentiment lexicon affects the classification effect. Therefore, the construction of sentiment lexicon is the foundation of the sentiment classification.

There are some different sentiment lexicon resource available in Chinese language, such as HowNet [7], National Taiwan University Sentiment Dictionary (NTUSD), etc. In this paper we adopt HowNet and NTUSD as the basis sentiment lexicon. Considering the importance of network language, we manually collected some words to extend the basis sentiment lexicon. The actual number of basis sentiment lexicon is shown in following Table 1:

Table 1. The basis sentiment lexicon.

Lexicon	Positive words	Negative words
HowNet	3786	3908
NTUSD	2510	6846
Network language	156	208
Total	10835	159388

Negation word is the words that can change the sentiment polarity of sentence. Degree adverbs usually appear in front of the modified sentiment words and can enhance or weaken the emotional intensity of sentiment words. Some of the negation and degree adverbs are shown in Table 2.

Table 2. The result of sentiment polarity analysis for words.

Negative words		Degree adverbs words	
Number	Words	Number	Words
1	No	1	Especially
2	Do not	2	Extremely
3	None	3	A little
4	Never	4	More
5	Neither	5	Very much

Most sentiment analysis technology only consider sentiment words, but in Sina microblogging, people usually use emoticons to express their sentiment, especially in young groups. In this paper we use the emotions provided by Sina microblogging platform and divide these emoticons into positive and negative.

3.2 Word Embedding

Lexicon-based method can achieve sentiment classification, but it can not capture the contextual cues and deeper text information, so its accuracy is not high. However, word embedding can dig the relationship between words, such as synonyms. At present, word embedding can fully explore the "synonyms", such as "die" and "pass away", but can not deal with "word polysemy", such as "Apple" (both can say Apple phone, computer and fruit). Word2Vec as a word embedding-based method can get the vector representation of each word of the text whose sentiment polarity we need to predict. Word2Vec uses a three-layer neural network [8] to obtain a representation of a word in a vector space. After getting word vectors, in order to improve the performance of our hybrid method we need to select more effective features using lexicon constructed in Sect. 2. We define some patterns described in Table 3 to extract effective features. Because

Table 3. Feature patterns.

Pattern	Example
Sentiment words	Happy
Negation words + sentiment words	Not happy
Degree adverbs + sentiment words	Very happy
Negation words + degree adverbs + sentiment words	Not very happy
Degree adverbs + negation words + sentiment words	Especially not happy

of the randomness of the position of emoticon, we do not use these patterns to extract emoticon features. We compare the words in the corpus with emoticons, adding the matching words to the feature set.

3.3 Clustering

In Sect. 3.2 we have obtained the vector representation of words that related to sentiment of text. In order to achieve our purpose of sentiment analysis we need a classifier to determine the sentiment polarity of the text. A large number of studies have shown that SVM as a classifier has better performance than other state-of-art models and has been wildly used because of its excellent generalization ability (not easy to fit) [9]. Considering the above conclusion we utilize SVM to classify the text. SVM is a supervised classification algorithm and needs a lot of data and the label corresponding to data to train an effective model. Firstly, we construct SVM classifier. Then the text vector of training data is applied to train the classifier. Finally, the testing data is tested on the classifier.

4 Experiments

In this section, in order to prove the effectiveness of our hybrid method (SAFCM) a series of experiments were done. We apply our method on two different corpus and compare with three other methods. All of the experiments in this paper are implemented in python language.

4.1 Data Sets

In order to ensure the fairness and effectiveness of the experiment, we apply our method on two public datasets, one is Sina micro- blog sentiment corpus (SMSC), the other is the corpus from The Fifth Chinese Opinion Analysis Evaluation (COAE2014).

4.2 Experiment Results

In terms of performance evaluation, we compare our hybrid method (SAFCM) with other three algorithms (lexicon-based method, word2vec-based method and SVM-based method) to prove the effectiveness of our method. We compare their precision rate, recall rate and F1-SCORE, respectively. The comparative results are shown in Fig. 2a–c. Comparison results show that SAFCM is better than other methods in terms of precision, recall and F1-SCORE, at the same time the results also proved the effectiveness of the proposed hybrid method in this paper.

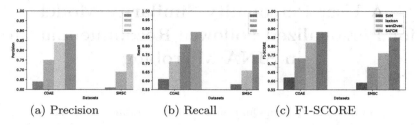

(a) Precision   (b) Recall   (c) F1-SCORE

Fig. 2. The comparsion between SAFCM and other methods in terms of precision, recall and F1-SCORE.

5   Conclusion and Future Work

Because of the importance of the sentiment analysis, it received increasing attention of scholars from diverse fields in recent years. In this paper, different from traditional method we combine sentiment lexicon, word embedding and SVM to deal with sentiment analysis problem for Chinese microblog. The experiment results show the effectiveness of our hybrid method. But we only use static lexicon without dynamically updating lexicon. In our future work, we will overcome this insufficient.

Acknowledgment. This research is supported by the National Natural Science Foundation of China (Grant nos. 61472329 and 61532009), youth fund of China (no. 61602389) and the Innovation Fund of Postgraduate, Xihua University (no. ycjj201671).

References

1. Pang, B., Lee, L., Vaithyanathan, S.: Thumbs up? Sentiment classification using machine learning techniques. In: ACL-2002 Conference on Empirical Methods in Natural Language Processing, pp. 79–86 (2002)
2. Shen, Y., Li, S., Zheng, L., Ren, X., Cheng, X.: Emotion mining research on microblog. In: IEEE Symposium on Web Society (SWS 2009), pp. 71–75 (2009)
3. Hung, C., Chen, S.J.: Word sense disambiguation based sentiment lexicons for sentiment classification. Knowl. Based Syst. **110**, 224–232 (2016)
4. Tripathy, A., Agrawal, A., Rath, S.K.: Classification of sentiment reviews using n-gram machine learning approach. Knowl. Based Syst. **57**, 117–126 (2016)
5. Katz, G., Ofek, N., Shapira, B.: ConSent: context-based sentiment analysis. Knowl. Based Syst. **84**, 162–178 (2015)
6. Turney, P.D.: Thumbs up or thumbs down? Semantic orientation applied to unsupervised classification of reviews. In: Proceedings of Annual Meeting of the Association for Computational Linguistics, pp. 417–424 (2002)
7. Dai, L., Liu, B., Xia, Y., Wu, S.K.: Measuring semantic similarity between words using HowNet. In: International Conference on Computer Science and Information Technology, pp. 601–605 (2008)
8. Bengio, Y., Ducharme, R., Vincent, P., Janvin, C.: A neural probabilistic language model. J. Mach. Learn. Res. **3**, 1137–1155 (2003)
9. Tang, H., Tan, S., Cheng, X.: A survey on sentiment detection of reviews. Expert Syst. Appl. **36**, 10760–10773 (2009)

A User Personality-Similarity Model for Personalized Followee Recommendation in SINA Microblog

Pan Xiao$^{(\boxtimes)}$, YongQuan Fan, YaJun Du, and Chun Yu

School of Computer and Software Engineering, Xihua University,
Chengdu 610039, China
1049434055@qq.com

Abstract. Followee recommendation plays an important role in information sharing over microblogging platforms. But as the popularity of microblogging sites increases, the difficulty of deciding who to follow also increases. To solve this problem, in this paper, we propose a User Personality-Similarity (UPS) model for followee recommendation, a novel personality followee recommendation scheme over microblogging systems based on user attributes and the big-five personality model. It quantitatively analyses the effects of user personality in followee selection by combining personality traits with the most commonly user-based predictive factors in microblog. We conduct comprehensive experiments on a large-scale dataset collected from Sina Weibo, the most popular mircoblogging system in China. The results show that our scheme greatly outperforms existing schemes in terms of precision and an accurate appreciation of this model tied to a quantitative analysis of personality is crucial for potential followees selection, and thus, enhance recommendation.

Keywords: Followee recommendation · Microblog
Personality traits · User similarity

1 Introduction

Since the emergence of social networks and microblogging sites, such as Twitter and Sina Weibo, hundreds of millions of users have become to use the microblogging service as a tool to propagate and share information on the Internet [1]. Personality affects everyone's thinking, behavior and decision-making and maintain its long-term stability and affect its social relations [2]. The behavior of users in the online world and in the real world is the same [3]. However, there are few researches on personality theory in the previous studies.

This work is supported by the National Nature Science Foundation (Grant No. 61472329 and 61532009), the Key Natural Science Foundation of Xihua University (Z1412620) and the Innovation Fund of Postgraduate, Xihua University.

J. Li et al. (Eds.): CCKS 2017, CCIS 784, pp. 158–164, 2017.
https://doi.org/10.1007/978-981-10-7359-5_16

The researcher combines personality psychology and social network analysis to analyze and predict the user's personality through the user's behavior and other data. The potential value of personality prediction will be very helpful in solving these problems.

The rest of this paper is organised as follows. In Sect. 2, we discuss related work. Section 3 presents the followee recommendation method and the UPS model we propose. In Sect. 4, we evaluate the performance of our design experiment, present the proposed weights for quantitatively evaluating users' personality and how to combine it with other recommendation and compare to existing schemes. Section 5 summarises the conclusions obtained from the performed experiment evaluation.

2 Related Work

Personalized recommendation technology is the core and critical technology of E-commerce recommendation system. However, these traditional methods only focus on how to improve the accuracy of recommendation, while ignoring the inherent characteristics of the user behavior is determined by their personality characteristics [4].

Some more related worked on followee recommendation has also been proposed. Most of the existing followee recommendation systems on micro-blogging platforms rely on either topological or content-based factors. Hannon et al. [5] use bag-of-word model to exploit the content information created by a user and recommend him/her followees based on the content similarity between the candidate followees and the target user. Sun et al. [6] construct a diffusion graph to select a small subset of tweets as recommended emergency new feeds for regular users. There are a few people use personality trait in followee recommendation systems. Wu et al. [7] generate personalized tags for Twitter users to label their interest by extracting keywords from tweets they post. Armentano et al. [8] propose a followee recommender system using social relation features, including user popularity and number of common friends, to measure the relevance among users in Twitter. However, the recommendation effect of adding personality traits has improved compared with the traditional recommendation method, but there is much room for improvement.

3 Our Method

3.1 User-Based Factor

The Analysis of User Attributes of Microblogging. In this section, we discuss the representation of the micro-blogging user model. In this work, we consider 4 properties of user information, which are location information, micro-blogging text, social information and interactive information. Thus, the user model of the user u is represented as follows:

$$Profile_{user}(u) = \{Place(u), Posts(u), Relation(u), Interaction(u)\} \quad (1)$$

where Place(u) is the location information of the user u, both of which are short text and can be represented as a string. Post(u) represents the long text that the user u has released into the micro-blogging, which is represented as a text vector. Relation(u) represents the social information of u, including two kinds of attribute information, which is concerned about attention information, fans information. Interaction(u) represents the interactive information of user u, including two kinds of attribute information, that is, forwarding information and comments.

Firstly, we define the user(u, v) in the various attribute similarity and then weighted and combined. The similarity is given by the following equation:

$$
\begin{aligned}
sim(u,v) = {} & \omega_1 sim(Place(u), Place(v)) + \omega_2 sim(Post(u), Post(v)) \\
& + \omega_3 sim(Relation(u), Relation(v)) \\
& + \omega_4 sim(Interaction(u), Interaction(v))
\end{aligned} \quad (2)
$$

In essence, ω_i is the weight of each attribute similarity, and $\omega_1 + \omega_2 + \omega_3 + \omega_4 = 1$. The sim(Place(u), Place(v)) = 1 if the information of the provincial and city of the Place(u) and Place(v) are the same. The sim(Place(u), Place(v)) = 2/3 if the information of provincial of the Place(u) and Place(v) is the same, but the city information is different. For two users(u) and user(v), their micro-blogging text can be represented as two text feature vectors: $Post(u) = (w_{u1}, w_{u2}, ..., w_{un}), Post(v) = (w_{v1}, w_{v2}, ..., w_{vn})$. The text similarity $sim(Post(u), Post(v))$ is calculated by cosine similarity, which used TF-IDF. The similarity of the relationship information between the two users are computed weighted averages of $Followee(u)$ and $Followee(v)$. Also, we use forwarding and comment numbers to compute the similarity of $sim(Interaction(u), Interaction(v))$.

3.2 Personality-Based Factors

The Big-Five Personality Model. The "Big Five" model of personality dimensions has emerged as one of the most well-researched and well-regarded measures of personality structure in recent years. The model's five domains of personality, Openness, Conscientiousness, Extroversion, Agreeableness, and Neuroticism, were conceived by Tupes and Christal as the fundamental traits that emerged from analyses of previous personality across age, gender, and cultral lines.

The Matching Calculation of the Personality Traits. TextMind is a Chinese language psychological analysis system developed by Computational Cyber-Psychology Lab, Institute of Psychology, Chinese Academy of Sciences. Through the relationship between the function of these words and the text, in this paper, we can obtain the relationship between each word in TextMind Chinese psychological analysis system dictionary and each specific factor in the big five personality. The character factor values for the ith dimension in word w are defined as follows:

$$BFM(\omega_i) = \frac{\sum_{j=1}^{n} P_{ij}}{n} \tag{3}$$

In Eq. (3), ω_i denotes the ith personality factor in word w. n indicates that word w has n functional nouns. P_{ij} represents the relationship between the jth word function and the ith personality factor. The functional word of TextMind and BFM corresponding a correspondence table can be divided into five groups.

In this paper, we calculate the personality score vector of each dimension of the big five personality according to the BFM correspondence table [9]. The user u's personality score in the ith dimension is calculated as follows:

$$Score(u)_i = \frac{\sum_{j=1}^{N_i} k_{\omega_j} \cdot BFM(\omega_{ji})}{N_i} \tag{4}$$

Where $BFM(\omega_{ji})$ denotes the personality factor value of the jth-class word in the ith dimension in the micro-blogging text published by the user u. k_{ω_j} represents the word frequency of the jth-class word ω of the microblogging text published by the user u. N_i is the total number of functional words that are statistically relevant under the ith personality dimension. We use the Eq. (4) to calculate the average score between five dimensions in the table, which were μ_E, μ_A, μ_C, μ_N, μ_O, and then we put the user scores compared with the average correlation score.

The User Personality-Similarity Model. In this Section, we proposed the user personality-similarity model to make the final recommendation. The personality matching score between a user u and the potential blogger pf is expressed as follows:

$$TPM(u, pf) = \mu(\sum MS(u, pf, dim)) \tag{5}$$

In Eq. (5), $TPM(u, pf)$ is the total personality matching (TPM) score between user u and potential followee pf. μ is the average value of each dimension. $MS(u, pf, dim)$ indicates the personality matching score of the user u and the potential recommendation followee pf in a certain dimension.

Next, the matching score calculation formula for each dimension is defined as follows [10]:

$$scoreAgreement(u, pf, dim) = \begin{cases} 0.5 & \text{both u and pf are dimension} \\ 0.25 & \text{either u or pf are dimension} \\ 0 & \text{None is dimension} \end{cases} \tag{6}$$

Finally, we formally define the evaluation of the formula in the UPS model as follows,

$$FScore = \gamma_1 sim(u, v) + \gamma_2 TPM(u, pf) \tag{7}$$

Where γ_i is the weight value of the formula, and $\gamma_1 + \gamma_2 = 1$. Finally, we use this formula as the final evaluation formula in followee recommendation.

4   Experiment

In this experiment, We use the microblogging users between the various attributes of information similarity calculation method for user recommendation experiment. In this paper, two evaluation indicators are used to evaluate the user similarity algorithm which are p@N and average precision.

We use the large-scale dataset crawled from Sina Weibo, the most popular microblogging system in China. The dataset contains 256.7 million users' social link information and 550 million tweets. The dataset include tweets, user relations and user background information.

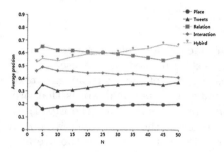

Fig. 1. Recommendation results based on user similarity

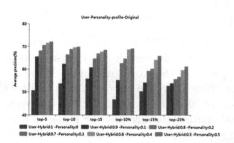

Fig. 2. Personality-aware of compared methods

4.1   Results

In the experiment, the weight of the attribute similarity calculation formula between the user(u) and user(v) is calculated by Analytic Hierarchy Process (AHP).

Figure 1 shows the experimental results of the above similarity on P@N, which reflects the motivation of the experiment is to examine the user information and its four kinds of attribute information (place, tweets, relation, interaction) in the calculation of user similarity performance. The experiment results show that compared with the four attribute information of user information, the P@N results of the similarity of relation are best in the previous P@25 and the overall similarity of the users is the second.

Unlike when computing the personality profiles in which ever term appearing in tweets was considered, terms in tweets were filtered according to a processing approach. The approach considered the full-text of tweets (named Original). The effect of adding personality as a factor in user-based followee recommendation is shown in Fig. 2. The figure summarises the average precision for each of the predefined N, including results for six linear combinations of factor's weights. As a result, it can be stated that considering a quantitative analysis of personality in combination with user-based factors could help to correctly place the most important or interesting users in the first positions of the ranking of suggested users.

5   Conclusions

In this paper, we propose a User Personality-Similarity (UPS) model for followee recommendation, a novel personality followee recommendation scheme over microblogging systems based on user attribute and the big-five personality model. This paper analysed how user personality conditions the followee selection process by combining a quantitative analysis of personality traits with the most commonly used predictive factors for followee recommendation. The combined attributes were insert into a recommendation algorithm that computed the similarity among target users and potential followees. We conduct experiments using large-scale traces form Sina Weibo to evaluate our design. Results show that UPS model greatly outperforms existing recommendation schemes.

References

1. Dong, Y., Tang, J., Wu, S., et al.: Link prediction and recommendation across heterogeneous social networks. In: IEEE International Conference on Data Mining, pp. 181–190. IEEE (2013)
2. Selfhout, M., Burk, W., Branje, S., et al.: Emerging late adolescent friendship networks and Big Five personality traits: a social network approach. J. Personal. **78**(2), 509–538 (2010)
3. Barrick, M.R., Mount, M.K.: The Big Five personality dimensions and job performance: a meta-analysis. Pers. Psychol. **51**(1), 849–857 (1991)
4. Gao, R., Hao, B., Bai, S., et al.: Improving user profile with personality traits predicted from social media content. In: ACM Conference on Recommender Systems, pp. 355–358. ACM (2013)
5. Hannon, J., Mccarthy, K., Smyth, B.: The pursuit of happiness: searching for worthy followees on Twitter. In: Irish Conference on Artificial Intelligence and Cognitive Science, pp. 1761–1762 (2011)
6. Sun, A.R., Cheng, J., Zeng, D.D.: A novel recommendation framework for microblogging based on information diffusion. Social Science Electronic Publishing, vol. 12(7) (2010)
7. Wu, W., Zhang, B., Ostendorf, M.: Automatic generation of personalized annotation tags for twitter users. In: Proceedings of the Human Language Technologies: Conference of the North American Chapter of the Association of Computational Linguistics, DBLP 2010, Los Angeles, California, USA, 2–4 June 2010, pp. 689–692 (2010)
8. Armentano, M.G., Godoy, D.L., Amandi, A.: A topology-based approach for followees recommendation in Twitter. In: International Workshop on Intelligent Techniques for Web Personalization & Recommendation, pp. 22–29 (2013)

9. Mairesse, F., Walker, M.A., Mehl, M.R., et al.: Using linguistic cues for the automatic recognition of personality in conversation and text. J. Artif. Intell. Res. **30**(1), 457–500 (2007)
10. Tommasel, A., Corbellini, A., Godoy, D., et al.: Personality-aware followee recommendation algorithms: an empirical analysis. Eng. Appl. Artif. Intell. **51**(C), 24–36 (2016)

CrowdGeoKG: Crowdsourced Geo-Knowledge Graph

Jiaoyan Chen[1], Shumin Deng[2], and Huajun Chen[2(✉)]

[1] GIScience, Heidelberg University, Heidelberg, Germany
j.chen@uni-heidelberg.de
[2] College of Computer Science, Zhejiang University, Hangzhou, China
{231sm,huajunsir}@zju.edu.cn

Abstract. Geo-Knowledge which is known as semantic enriched geographic information plays an important role in many intelligent applications like named-entity recognition and information retrieval. With the development of Internet, volunteers on web-based crowdsourcing platforms like OpenStreetMap (OSM) and Wikidata have contributed big geographic data which however have not been widely studied towards extracting and linking geo-knowledge. In this paper, we presented a crowdsourced geographic knowledge graph named CrowdGeoKG which extracted different kinds of geo-entities from OpenStreetMap and enriched them with human geography knowledge from Wikidata. We further exploited the part of CrowdGeoKG in China, studying the linkage between OpenStreetMap geo-entities and Wikidata geo-entities. CrowdGeoKG is stored in both RDF (Resource Description Framework) and JSON-LD formats, and shared for re-usage on an open knowledge graph community named OpenKG.

Keywords: Crowdsourced geographic information
Knowledge graph · OpenStreetMap · Wikidata

1 Introduction

The knowledge graph is a knowledge base originally used by Google to enrich the semantics of search results from a wide variety of data sources, and now has been widely applied in different artificial intelligence applications including nature language understanding, knowledge engineering, information retrieval, question answering, and even predictive analytics. The geo-knowledge graph is the one created by structuring existing geospatial datasets into a semantic format with rich domain knowledge, and has become a promising support tool to many geospatial technical challenges like named-entity recognition and classification, toponym disambiguation and resolution, spatial footprints, spatial reasoning, user context, etc. [4]. Meanwhile, by linking different datasets, geo-knowledge graphs directly solve some real world problems like enriching street names for better information retrieval over maps [1] and mining correlation between geospatial variables and social indexes [10, 11].

© Springer Nature Singapore Pte Ltd. 2017
J. Li et al. (Eds.): CCKS 2017, CCIS 784, pp. 165–172, 2017.
https://doi.org/10.1007/978-981-10-7359-5_17

On the other hand, crowdsourced (volunteered) geographic information increases at a fast speed in recent years and is playing a more and more important role in many domains like humanitarian aid for disasters and smart urban planning. In the Haiti earthquake in 2010, most roads, damaged buildings and so on were added on OpenStreetMap (OSM) in a short time after the earthquake happened [8,12]. By July 2016, OSM had over 2.8 million accumulated registered users, 3.25 billion accumulated nodes, and 250 million accumulated ways[1]. Except for geographic wiki systems, other general purpose crowdsourcing websites like Wikipedia and Wikidata have also contributed a large number of location tagged pages. For example, by October 2015, 12.3% of the pages on Wikidata (over 1.9 million) are about administrative territorial entities[2]. Utilizing data from such crowdsourced geographic information systems provides a new approach for building a large scale geo-knowledge graph.

Currently, there are only a limited number of studies that aim at building geo-knowledge bases from crowdsourced geographic information. OSMonto is an ontology for OSM tags (e.g., (*building, yes*)), but does not contain any instances [5]. LinkedGeoData is an early project that mapped OSM data into RDF data model, interlinked the data with other datasets and made the data accessible to machines and humans by a faceted geo-data browser [2]. The OSM Semantic Network, which contains a machine-readable representation of OSM tags is another geo-knowledge base extracted from OSM Wiki website [3]. Both LinkedGeoData and the OSM Semantic Network extract geospatial data mainly from one single OSM data source. Although OSM tags contains a large number of key value pairs, their common sense knowledge about the geo-entities are limited in comparison with general purpose wiki websites. Yago2 [7] and Clinga [9] are two knowledge bases with human geography knowledge from Wikipedia and Baidu Baike respectively, but only cover a limited number of ground objects, especially for roads and point of interests (POIs) like buildings, stations and parks in urban areas.

In this paper, we propose to extract and integrate geographic information from both OSM and Wikidata for building a crowdsourced geo-knowledge graph called CrowdGeoKG. Geo-entities including three different kinds of OSM features, namely Node, Way and Relation are first extracted from a big OSM data repository, and then integrated with Wikidata pages with a union operation based on (i) the OSM tag whose key is *wikidata* and (ii) the Wikidata page property of *OpenStreetMap Relation identifier*. The integrated geospatial data are transformed into two Semantic Web formats RDF (Resource Description Framework) and JSON-LD [6] with a schema for knowledge representation and publish. We further exploited CrowdGeoKG by studying some of its statistics including the scale, the rate of linkage between OSM geo-entities and Wikidata geo-entities, the missing rate of some tags and so on. Compared with the existing geo-knowledge bases, our work contributes in the following aspects:

---

[1] http://wiki.openstreetmap.org/wiki/Stats.
[2] https://www.wikidata.org/wiki/Wikidata:Statistics.

- CrowdGeoKG contains rich geo-knowledge covering both human geography entities (e.g., administrative regions) and ground object entities (e.g., point of interests and roads),
- CrowdGeoKG is as far as we know the first work that exploits geo-knowledge from Wikidata and links them with geo-knowledge from OpenStreetMap.

The left of the paper is organized as follows. Section 2 introduces the technical details of the schema and the geo-knowledge extraction process. Section 3 presents the statistics of CrowdGeoKG. Section 4 concludes the paper and introduces our future work.

2   Technical Framework

2.1   Schema Design

For knowledge representation and publishing, a schema and some prefixes are designed for CrowdGeoKG, as shown in Fig. 1, where a geo-entity *cgk:2780071* from OSM and a geo-entity *cgk:Q71339* from Wikidata are used as two instance examples. The unique ID from OSM and Wikidata is used in naming a geo-entity with a prefix *cgk* designed for CrowdGeoKG. Geo-location (position information) of a geo-entity is represented in Well-Known Text (WKT) format which is a text markup language for representing vector geometry objects on map. Geometries of point, multi-point, line, multi-line, polygon and multi-polygon can all be represented by WKT format and visualized on most of the existing geographic information systems. The predicate of *geo:asWKT* is adopted.

Each geo-entity has a series of properties like *cgk:name_en* and *cgk:name_zh* for representing its basic semantic information. The complete semantics of a geo-entity are represented by an existing ontology named OSMonto [5] which includes hierarchical classes defined according to OSM tags. The geo-entity will be linked with the classes with the predicate of *rdf:type*. For example, if a geo-entity is labeled by a tag *(shop, car_repair)*, it will be an instance of the class *k_shop* and its subclass *v_car_repair* defined in OSMonto.

A mechanism is designed for integrating the geo-entities from OSM and Wikidata. A top class *cgk:GeoEntity* is used for representing all the geo-entities, and two disjoint subclasses named *cgk:OSMGeoEntity* and *cgk:WikiGeoEntity* are used for geo-entities extracted from OSM and Wikidata respectively. The class of *cgk:OSMGeoEntity* further includes three disjointed subclasses *cgk:OSMNode*, *cgk:OSMWay* and *cgk:OSMRelation* respectively for three different kinds of OSM features (i.e., ground objects), namely Node, Way and Relation. Node usually represents a ground geographic object in point format (e.g., points of interest and places), Way usually represents a ground geographic object in line or polygon format (e.g., roads and buildings), while Relation usually represents a ground geographic object in polygon or multi-polygon format (e.g., boundaries). The linkage between an OSM geo-entity and a Wikidata geo-entity is represented by the predicate of *sameAs*.

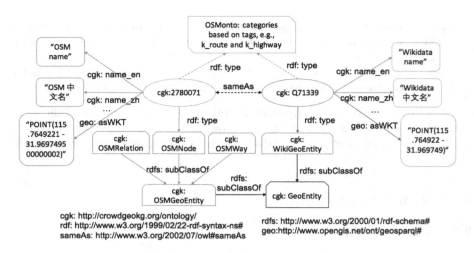

Fig. 1. The schema and prefixes of CrowdGeoKG.

2.2 Data Transformation and Linking

To construct CrowdGeoKG, geo-knowledge are (I) extracted from OSM and Wikidata, (II) interlinked according to the references to each other and (III) transformed into RDF and JSON-LD formats, as shown in Fig. 2.

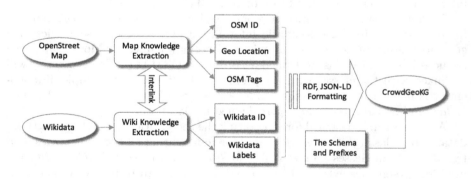

Fig. 2. The data flow of geo-knowledge extracting and geo-entity linking.

The latest OSM data of the whole world are available on Geofabrik's free download server[3] in XML format. The XML data includes a list of items each of which is an OSM feature (geo-entity) of Node, Way or Relation. As a Way item references Node items, and a Relation item references Node items and Way items, our program traversals the OSM data three times, dealing with Node

---

[3] http://download.geofabrik.de/.

items, Way items and Relation items in order. OSM id, geo-location and tags of each item are extracted. The dataset of Wikidata downloaded[4] is in JSON format. Similarly, Wikidata id, geo-location and the properties of each JSON item are extracted. To process the big raw OSM dataset and Wikidata dataset (e.g., the OSM dataset file of China exceeds 10G and the Wikidata dataset file exceeds 90G) on a single workstation, a cache strategy is adopted.

In extracting geo-location, the technical challenge lies in determining the geo-entity's geometry format. A Node item must be in the WKT format of *POINT* as it only contains a pair of latitude and longitude. A Way item is composed of a list of Node items. Its WKT format is (I) *POLYGON* if the referenced Node items form a closed loop, (II) *LINESTRING* elsewise. A Relation item includes three cases:

- When the Relation item is composed of only Node items, its geometry format is (I) *POINT* if there is only one Node item, and (II) *MULTIPOINT* if there are more than one Node items.
- When the Relation item is composed of only Way items, its geometry format is (I) the same as the Way item if there is only one Way item, (II) either *MULTILINESTRING* or *MULTIPOLYGON* if there are multiple Way items and the geometries that the Way items form are the same, and (III) *COLLECTION* if the geometries that its Way items form are different.
- When the Relation item is composed of both Way items and Node items, its geometry format must be *COLLECTION*.

To extract all the linked pairs of the OSM geo-entity and the Wikidata geo-entity, a union-based operation which scans the geo-entities once is adopted. If the geo-entity from OSM contains a tag with the key of *wikidata*, the value of the tag which is a Wikidata page id is used to find the Wikidata geo-entity that the OSM geo-entity should link. On the other hand, if the geo-entity from Wikidata contains a property of *OpenStreetMap Reference*, the value of the property which is an OSM feature id is used to find the OSM geo-entity that the Wikidata geo-entity should link.

3   Dataset Exploitation

To exploit CrowdGeoKG, we present (i) the scale of the knowledge graph, (ii) the linkage between OSM and Wikidata, and (iii) the knowledge missing of the geo-entities from OSM. The results are based on the part of CrowdGeoKG in China which is available on OpenKG[5].

As shown in Table 1, CrowdGeoKG contains over 5 million OSM geo-entities in China, among which the three classes *OSMNode*, *OSMWay* and *OSMRelation* account for 31.50%, 67.52% and 0.98% respectively. We can find that over 300 thousand of the geo-entities are linked to Wikidata pages (overall column),

---

[4] https://www.wikidata.org/wiki/Wikidata:Database_download.
[5] http://openkg.cn/dataset/crowdgeokg.

Table 1. Number of OSM geo-entities and the linkage to Wikidata geo-entities. Linkage rate is the percentage of OSM geo-entities that are linked to Wikidata geo-entities.

	OSMNode	OSMWay	OSMRelation	Overall
Geo-entities #	1,605,886	3,442,299	49,749	5,097,934
Wikidata linkage #	331,090	533	5,803	337,446
Linkage rate	20.62%	0.02%	11.66%	6.62%

accounting for 6.62% (linkage rate), which is actually quite low. Another interesting phenomenon is that the linkage rate varies a lot from geo-entities types. *OSMNode* geo-entities have as high linkage rate as 20.62%, while the linkage rate of *OSMWay* geo-entities is as low as 0.02%. This is because the type of the ground objects that Node, Way and Relation represent on OSM is biased, while most of geo-entities from Wikidata are about human geography.

We further study the types of OSM geo-entities that have linkage to Wikidata by exploiting their OSM tags, as shown in Fig. 3. Among the geo-entities of *OSMNode*, we can find over 70% are tagged by the key of *place*, and most of the other top-10 keys are *population* and address related keys. This means most of the *OSMNode* geo-entities that have linkage to Wikidata are places like village, town and city. Similarly, for *OSMWay*, most of the geo-entities that are linked with Wikidata are buildings, tourism sites and waterways, while for *OSMRelation*, administrative regions account for the most according to the keys of *admin_level* and *boundary*. This phenomenon can confirm the above statistics that *OSMNode* and *OSMRelation* have much higher linkage rate than *OSMWay*, because places and administrative regions are more likely to be human geography and have Wikidata page.

By the way, the key of *wikipedia* which represents a linkage to a page of Wikipedia, another very popular wiki website ranks second for *OSMNode* and first for both *OSMWay* and *OSMRelation*. This means the linkage of OSM geo-entities to Wikidata is positively correlated with its linkage to Wikipedia.

By linking OSM geo-entities with Wikidata geo-entities, the semantics are enriched. For example, when the OSM geo-entity of *Node: Hangzhou (244080543)* is linked with the Wikidata geo-entity of *Q4970, Node: Hangzhou (244080543)* is enriched with knowledge like administrative territorial entities, sister cities, local dialing, elevation above sea level and so on. Figure 4 shows that a large part of the geo-entities of *OSMNode*, *OSMWay* and *OSMRelation* miss the labeling of English name (i.e., the key of *name:en*) or Chinese name (i.e., the key of *name:zh*). Totally, over 70% of the OSM geo-entities miss the labeling of *name:zh*, while the missing rate of *name:en* exceeds 50%. These missed Chinese name and English name information however is fully complemented by the linked Wikidata geo-entities. Enriching the OSM geo-entities with Chinese name makes CrowdGeoKG much more practical for applications in China.

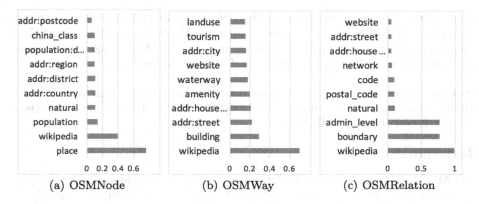

Fig. 3. Top-10 keys of the tags of the OSM geo-entities that have linkage to Wikidata geo-entities. The keys for name (e.g., *name* and *name:zh*) and the keys for meta information (e.g., *source*) are ignored.

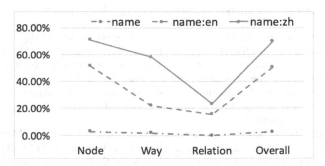

Fig. 4. The missing rate of three OSM tag keys (*name*, *name:en* and *name:zh*).

4 Conclusion and Future Work

This paper presented our crowdsourced geo-knowledge graph named Crowd-GeoKG. Geo-entities of *OSMNode*, *OSMWay* and *OSMRelation* were first extracted from OpenStreetMap together with the semantics from tags. They were then linked with geo-entities extracted from Wikidata with richer knowledge contributed by volunteers, especially those about human geography. These geo-knowledge were finally transformed into RDF and JSON-LD formats for re-using and publishing with a carefully designed schema. Different from the existing geographic knowledge bases, CrowdGeoKG is as far as we know the first work to exploit geo-knowledge in Wikidata and enrich knowledge of ground objects with general human geography knowledge. According to our statistics, Wikidata complements a large part of missed geo-knowledge in OSM, such as an geo-entity's Chinese name.

In the future, we will extend CrowdGeoKG in two aspects. First, a more general knowledge representation and reasoning framework, not only for static geographic information from crowdsourcing websites, but also for dynamic location-based data generated by citizen sensors and physical sensors will be designed. Air pollution data, traffic data and social web data will be enriched by the semantics and integrated. Second, prediction of the linkage between geo-entities will be studied with inference and learning technologies.

Acknowledgments. This work is funded by the Alibaba-ZJU joint project on e-Business Knowledge Graph and NSFC 61473260/61673338/61672393, and the Klaus Tschira Foundation (KTS) Heidelberg.

References

1. Almeida, P.D., Rocha, J.G., Ballatore, A., Zipf, A.: Where the streets have known names. In: Gervasi, O., et al. (eds.) ICCSA 2016. LNCS, vol. 9789, pp. 1–12. Springer, Cham (2016). https://doi.org/10.1007/978-3-319-42089-9_1
2. Auer, S., Lehmann, J., Hellmann, S.: LinkedGeoData: adding a spatial dimension to the web of data. In: Bernstein, A., Karger, D.R., Heath, T., Feigenbaum, L., Maynard, D., Motta, E., Thirunarayan, K. (eds.) ISWC 2009. LNCS, vol. 5823, pp. 731–746. Springer, Heidelberg (2009). https://doi.org/10.1007/978-3-642-04930-9_46
3. Ballatore, A., Bertolotto, M., Wilson, D.C.: Geographic knowledge extraction and semantic similarity in OpenStreetMap. Knowl. Inf. Syst. 37(1), 61–81 (2013)
4. Ballatore, A., Wilson, D.C., Bertolotto, M.: A survey of volunteered open geo-knowledge bases in the semantic web. In: Pasi, G., Bordogna, G., Jain, L. (eds.) Quality Issues in the Management of Web Information. Intelligent Systems Reference Library, vol. 50, pp. 93–120. Springer, Heidelberg (2013). https://doi.org/10.1007/978-3-642-37688-7_5
5. Codescu, M., Horsinka, G., Kutz, O., Mossakowski, T., Rau, R.: OSMonto-an ontology of OpenStreetMap tags. In: State of the map Europe (SOTM-EU) (2011)
6. W. W. W. Consortium et al.: JSON-LD 1.0: A JSON-based Serialization for Linked Data (2014)
7. Hoffart, J., Suchanek, F.M., Berberich, K., Weikum, G.: YAGO2: a spatially and temporally enhanced knowledge base from wikipedia. Artif. Intell. **194**, 28–61 (2013)
8. Horita, F.E.A., Degrossi, L.C., de Assis, L.F.G., Zipf, A., de Albuquerque, J.P.: The use of Volunteered Geographic Information (VGI) and crowdsourcing in disaster management: a systematic literature review (2013)
9. Hu, W., Li, H., Sun, Z., Qian, X., Xue, L., Cao, E., Qu, Y.: Clinga: bringing Chinese physical and human geography in Linked Open Data. In: Groth, P., Simperl, E., Gray, A., Sabou, M., Krötzsch, M., Lecue, F., Flöck, F., Gil, Y. (eds.) ISWC 2016. LNCS, vol. 9982, pp. 104–112. Springer, Cham (2016). https://doi.org/10.1007/978-3-319-46547-0_11
10. Ristoski, P., Bizer, C., Paulheim, H.: Mining the web of linked data with rapid-miner. Web Semant. Sci. Serv. Agents World Wide Web **35**, 142–151 (2015)
11. Vilches-Blázquez, L.M., Villazón-Terrazas, B., Saquicela, V., de León, A., Corcho, O., Gómez-Pérez, A.: GeoLinked Data and INSPIRE through an application case. In: Proceedings of the 18th SIGSPATIAL International Conference on Advances in Geographic Information Systems, pp. 446–449. ACM (2010)
12. Zook, M., Graham, M., Shelton, T., Gorman, S.: Volunteered geographic information and crowdsourcing disaster relief: a case study of the Haitian earthquake. World Med. Health Policy **2**(2), 7–33 (2010)

Author Index